NETL Accomplishments

2011

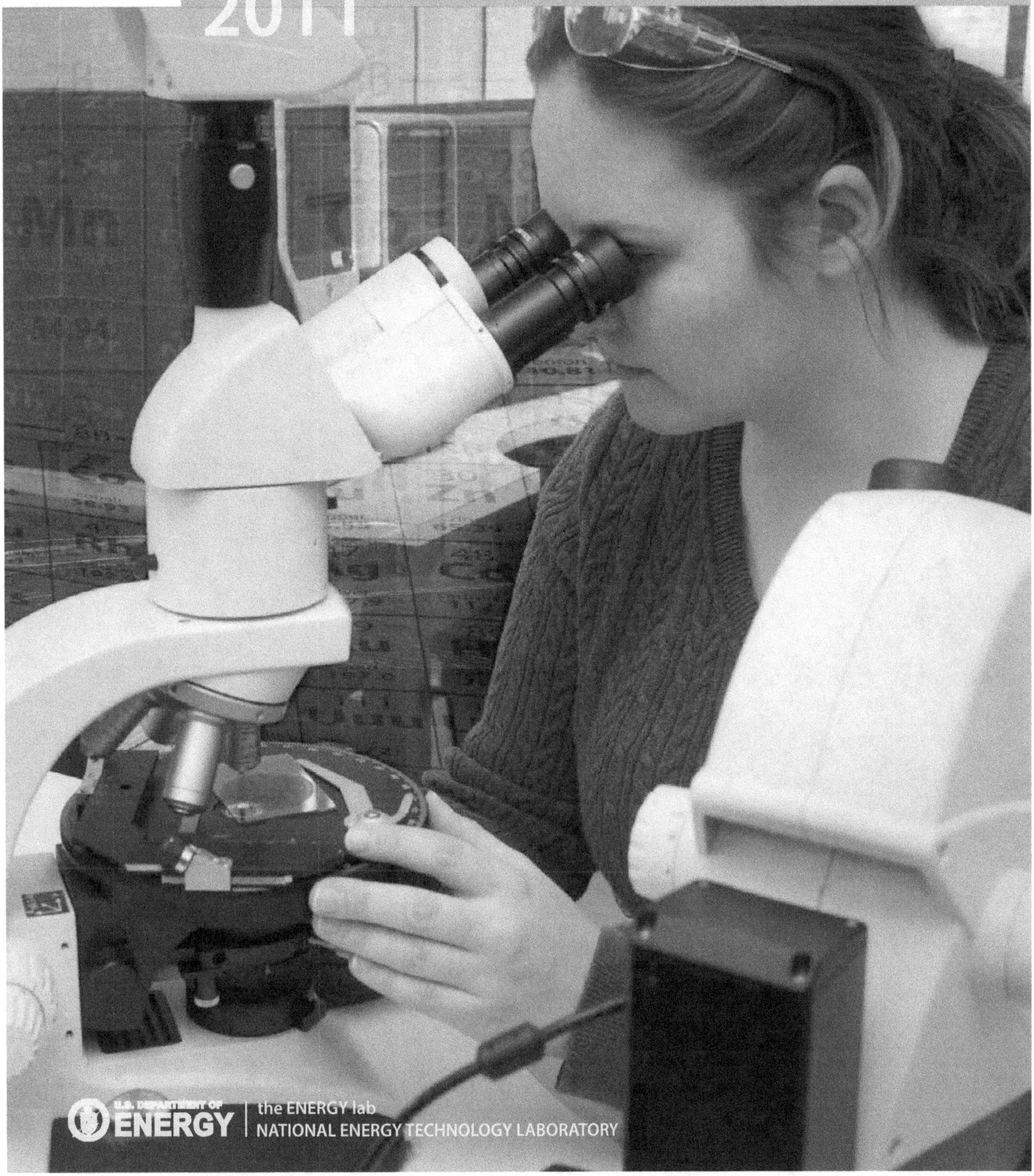

the ENERGY lab
NATIONAL ENERGY TECHNOLOGY LABORATORY

Mission

Advancing energy options to fuel our economy, strengthen our security, and improve our environment.

Cover image: NETL researcher Corinne Disenhof examines a basalt thin section under a geoscience laboratory petrographic microscope. NETL is investigating the effects of microbes on basalt during carbon sequestration, and petrography is one of several analysis methods being used. Others include scanning electron microscopy and x-ray diffraction.

Contents

Letter from the Director

Without question, the first quarter of the 21st century is an exciting time in the world of energy. The nation's push to develop sound energy policy is strong. The need to curb CO_2 emissions is urgent. The call for clean, affordable power is clear.

This is also an exciting time for energy researchers. The challenges we strive to overcome require far-reaching solutions: designing and building next-generation, near-zero-emission power plants; capturing, utilizing, and storing manmade carbon emissions; maximizing efficiencies; and getting the most from our energy resources with the least impact on our air, land, and waterways.

At the National Energy Technology Laboratory (NETL), we are confident in our ability to help America meet 21st-century challenges linked to fossil fuel recovery and use. After all, we have been meeting such challenges for generations. From mine safety to synthetic fuels, acid rain to mercury control, enhanced oil production to unconventional natural gas recovery, the legacy of public benefits realized by our programs stands as a testimony to the importance of our research and the careful means by which we steward federal investment in those programs.

Today, we continue to address complex energy issues—reducing the environmental impact of shale gas recovery, mitigating carbon emissions,

utilizing CO_2 to enhance our nation's oil and natural gas resources, and more. Our research is making a fundamental difference to the welfare of our nation. It also gives NETL and our stakeholders an opportunity to enhance our citizens' quality of life for generations to come.

In 2011, NETL realized a host of technical accomplishments and received recognition for our contributions to the energy arena. A full review of those achievements is available in the 2011 NETL Accomplishments, which contains the realized and estimated benefits provided by the programs implemented by our lab through the Office of Fossil Energy.

Thank you for taking time to browse our successes. I am proud of NETL's achievements and of the researchers and professional staff who make them possible.

Anthony V. Cugini, Director
National Energy Technology Laboratory

Advanced Power Systems

At NETL, projects are underway to develop technologies for use in generating new, advanced turbines that operate cleanly and efficiently using fuels derived from coal and containing high amounts of hydrogen.

Research into novel, next-generation energy technologies benefits modern energy production.

Reliable energy fuels the high quality of life that Americans enjoy. In conjunction, responsible environmental stewardship has emerged as an important component of power production and use. NETL researches technologies to enable more affordable, more efficient, and cleaner power generation. Our innovative coal research is economizing CO_2 capture at integrated gasification combined cycle power plants. We're developing technologies to capture coal combustion pollutants and finding ways to improve the efficiency and security of electricity transmission. Our scientists are also creating advanced materials that perform well in the harsh conditions of advanced power plants. Together with the Solid State Energy Conversion Alliance, NETL researches low-cost, high-performance fuel cells. And, in turbines research, NETL and its partners are developing and evaluating high-performance, low-emission concepts for cleaner, more efficient, and lower-cost power production.

NETL Researchers among Best in Industry for 2011

Dept. of Energy Honors NETL Researchers

The Secretary's Achievement Award is one of three types of DOE Secretary's Honor Awards given each year to DOE employees and contractors for outstanding service to the Department and the nation.

The Secretary of Energy Achievement Awards are the Department of Energy's highest internal, nonmonetary award a team can receive, and NETL is among the nine teams awarded, earning not one, but two awards. Recognized as DOE's "Academy Awards," the October 27 ceremony honored DOE employees from across the country for their outstanding work in the energy arena.

NETL was awarded for contributions made to two significant efforts in 2010: our nation's response to the Deepwater Horizon oil spill in the Gulf of Mexico and DOE's remediation activities at the Hanford nuclear materials production site in Washington State. In both challenges, researchers invested their time, effort, and expertise to generate solutions for critical problems facing the United States.

Members of the Flow Rate Technical Group/Nodal Analysis Team pose with Secretary Chu after receiving their awards.

Deepwater Horizon Flow Estimation Group/ Nodal Analysis Team—On April 20, 2010, the United States was struck with its worst environmental disaster. A blowout by BP's Deepwater Horizon well platform claimed the lives of 11 men, destroyed the drilling rig, and allowed an estimated 4.8 million barrels of oil to flow in the Gulf of Mexico before the well was successfully capped 87 days later.

NETL researchers were instrumental in ending the crisis. With more than 50,000 barrels of oil per day flowing into the Gulf of Mexico, determining the flow rate was the first step in identifying options for capping the well.

NETL's Dr. George Guthrie led the Nodal Analysis Team, coordinating the efforts of 46 scientists and team members from 6 DOE national labs—NETL, Los Alamos National Laboratory, Lawrence Berkeley National Laboratory, Lawrence Livermore National Laboratory, Pacific Northwest National Laboratory, and Oak Ridge National Laboratory—and the National Institute of Standards and Technology.

DOE's researchers employed state-of-the-art technologies and methods to develop unique approaches for solving the unprecedented problem. Dr. Guthrie's team worked to assemble, reduce, and analyze data and vet results with other professionals. They conducted their activities and formed analytical products quickly, helping to speed the ultimate solution and reduce the environmental cost of the disaster. Many of the processes developed for this task will now be refined and documented to guide the work of response analysts in the future.

The NETL federal and contractor team members who worked on the Nodal Analysis Team were George Guthrie, Grant Bromhal, Brian Anderson, Robert Enick, Roy Long, Shahab Mohaghegh, Bryan Morreale, Neal Sams, and Doug Wyatt.

Members of the Hanford Site V&V Evaluation Team with Secretary Chu.

The Hanford Site V&V Evaluation—The Hanford Site in Washington holds 53 million gallons of radioactive waste that is a byproduct of World War II and Cold War Era nuclear weapons production, as well as the production of commercial nuclear energy. Bechtel National, Inc., (BNI) is designing the Hanford Waste Treatment and Immobilization Plant to vitrify this waste into a stable glass form for safe, permanent storage.

BNI's plant design incorporates a pulse-jet system to maintain an even distribution of liquids and solids within the waste, minimizing the threat of nuclear reaction. Before proceeding to the build phase, BNI is conducting verification and validation (V&V) of the system using computational fluid dynamics to predict behaviors in the liquid-solid mix.

However, computational fluid dynamics presents a double-edged sword: the promise of accurate predictions that will optimize a system's design, and the risk of mischaracterization that could introduce error, or even danger, into a system.

DOE's Office of River Protection asked NETL to independently review and provide technical recommendations for BNI's V&V plan. Tom O'Brien and John VanOsdol, career fluid dynamics scientists at NETL, were asked to evaluate BNI's computer models, review the completed V&V effort, and provide feedback on whether BNI's computational fluid dynamics analyses would sufficiently confirm the validity of the pulse-jet mixer vessel design.

Dr. O'Brien and Dr. VanOsdol drew on the expertise of fellow NETL scientists Sofiane Benyahia, Madhava Syamlal, Tingwen Li, and Mehrdad Shahnam. They also involved two of the nation's top fluid dynamics scientists, Prof. Ismail Celik of West Virginia University and Dr. Urmila Ghia of the University of Cincinnati.

The team considered dozens of potential computational fluid dynamics approaches, and NETL made recommendations for improving the V&V, underscoring it with experimental data.

Based on the team's knowledge of power-production systems, in which fluid dynamics play a significant role, NETL also made recommendations on the pulse-jet mixer design, particularly in relation to erosion and segregation and concentration of particles. Although this analysis exceeded the original scope of the work, NETL researchers felt that the level of certainty necessary to mitigate risk at the treatment plant demanded that all findings be presented.

NETL is proud that the expertise and diligence of its researchers could aid the nation during these times of crisis and uncertainty.

Coal

NETL's coal research—advanced materials for ultra-efficient power generation, Smart Grid technologies, sorbents that capture pollutants from coal-fired plants, and gasification that can turn low-value feedstocks into energy or products and enable the capture of up to 90 percent of the CO_2 produced by coal-fired power plants—is helping to change the reputation of our most abundant fossil energy resource.

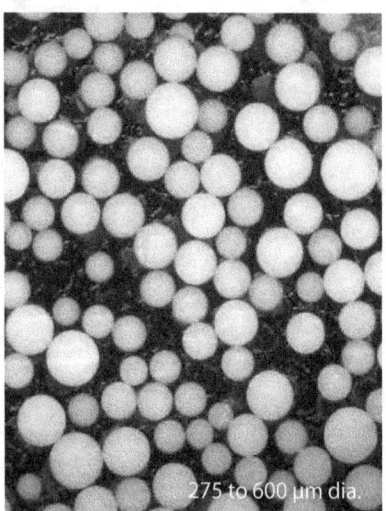

275 to 600 μm dia.

NETL won an FLC award for its technology transfer efforts related to the CO_2-capturing BIAS process. This image shows one of several sorbents that are used during the process.

NETL Sorbents Exhibit High Capacities for Arsenic Capture—

Collaborative research between NETL and Johnson Matthey has determined that NETL palladium-on-alumina sorbents possess extraordinarily high capacities for arsenic capture. Described in the October 2011 issue of *Fuel* (Vol. 90, No. 10), the work is timely, given the U.S. Environmental Protection Agency's 2011 announcement of new regulations for mercury and arsenic emissions from coal-utilizing facilities. The palladium sorbents were found to remove arsine (a gaseous form of arsenic) from simulated fuel gases at 204 °C and 288 °C —an outstanding result, particularly when considering that saturation had not likely been reached. Removing mercury and trace elements by sorbents at elevated temperatures preserves the high thermal efficiency of the integrated gasification combined cycle plant. During four extended exposures at a pilot gasification facility, the NETL sorbents removed virtually all of the mercury, arsenic, and selenium from synthesis gas slipstreams at 260–288 °C.

NETL Analyzes Water Consumption under Uncertainty for a Coal-Fired Power Plant—

By calculating optimized operating conditions (including reactor temperatures and pressures, reactant ratios and conditions, and steam flow rates and conditions) for a theoretical pulverized coal-fired power plant meeting environmental constraints, collaborators at NETL and the Vishwamitra Research Institute obtained reductions up to 6.3 percent for water consumption while allowing for uncertainty in air temperature and humidity. Described in the April 25, 2011, American Chemical Society publication *Environmental Science & Technology* (Vol. 45, No. 10), the study was performed using a novel "better optimization of nonlinear uncertain systems," or BONUS, algorithm, which dramatically decreased the computational requirements of stochastic optimization. Moreover, the methodology improved other plant performance parameters, such as gas emissions and cycle efficiency.

Small Business Innovation Research Yields Novel Technology for Membrane Fabrication—

With funding from an NETL-administered Small Business Innovation Research (SBIR) grant, product developers at Aegis Technology, Inc., produced two silver brazing alloys that exhibit excellent joining characteristics, wetting behaviors, microstructure, joining strength, and sealing capacity for ceramic-braze-ceramic assembly. Joint prototypes made using their novel reactive air-brazing technology demonstrated the advantages of the optimized braze compositions over the use of a conventional air braze formulation. While particularly suited for the hostile operating environments of energy conversion and power generation systems, the reliable, high-temperature sealing technology should have numerous applications wherever gas separation and purification is required for manufacturing, industrial, or medical purposes. Contact has been initiated with potential customers toward commercializing the technology for the gas separation device assembly.

NETL Applies for Patent on Novel Molten Hydroxide Gasification Process for Coal—

NETL researchers have invented a process for generating synthesis gas rich in hydrogen (about 80 percent) and methane (about 20 percent) at an intermediate range of gasifier temperatures (700 °C to about 1,000 °C) from coal or waste using alkali hydroxides and water. Proceeding in the absence of oxygen, the process converts alkali hydroxides into alkali carbonates, and all alkali species remain in the molten phase as catalysts for steam-gasification and methanation reactions. The alkali hydroxide species also capture sulfur and chlorine that may be in the feedstock, which gives the approach potential for converting high-sulfur coals or chlorine-containing wastes into synthesis gas to generate electricity with near-zero emissions of acid gases and greenhouse gases. Hydroxides consumed in gasification are regenerated outside of the gasifier from the alkali carbonates to produce nearly pure CO_2 for utilization or storage. While it is possible for the low-carbon and low-sulfur

synthesis gas leaving the gasifier to be sent directly to a gas turbine, the driving force for generating this high-methane synthesis gas is for fuel cell applications to reduce the amount of parasitic air cooling needed for solid oxide fuel cells (SOFCs). Integrating catalytic coal gasifiers with SOFCs has been shown to have fuel-to-electricity conversion efficiencies greater than 60 percent.

NETL Researchers Invent High-Efficiency Membrane Screening Tool—NETL scientists have designed a unique test system that accelerates the experimental determination of the performance of membrane materials by simultaneously characterizing up to 16 membrane films at a time. Based on conventional test system designs, the apparatus measures both membrane permeability and mixed gas selectivity. The new capability will allow researchers to develop novel CO_2 capture and oxygen separation technologies more rapidly than was possible before, including development of supported ionic liquid membranes and mixed matrix membranes. Rapid materials screening is critical to NETL's vertically integrated materials development approach, which relies on computational modeling to guide synthesis and fabrication activities but also requires quick feedback on performance to improve model quality.

Novel NETL Design Method Optimizes Efficiency and Flexibility of Integrated Gasification Combined Cycle Plants—Researchers at NETL and West Virginia University have developed through simulation a three-phase, top-down, optimization-based approach for designing integrated gasification combined cycle (IGCC) plants with pre-combustion CO_2 capture. In one iteration executing all three phases of the optimization approach, the optimization simulation yielded a net plant efficiency of 34.1 percent on a high-heating

value basis, compared to the 32.5 percent benchmark found in *Cost and Performance Baseline for Fossil Energy Plants*, Volume 1 (Rev. 2, 2010/1397). Described in the February 2, 2011, issue of the American Chemical Society publication, *Industrial & Engineering Chemistry Research* (Vol. 50, No. 3), the study identifies a number of key design variables and shows that the three-phase, top-down optimization approach is potentially useful for improving the efficiency and flexibility of IGCC power plants designed with CO_2 capture.

Using APECS v.2.0, design engineers can integrate, solve, and analyze co-simulations, such as for an IGCC power plant.

Newest Version of Advanced Process Engineering Co-simulator Reduces Time and Cost of Energy Process Innovations—The newest release of NETL's award-winning Advanced Process Engineering Co-simulator (APECS), version 2.0, has been integrated with the ANSYS® FLUENT® version 14.0 computational fluid dynamics (CFD) software to provide an improved engineering co-simulation tool suite that optimizes power generation processes. APECS version 2.0 further reduces the time and cost needed to foster energy plant innovations by seamlessly integrating process simulation with high-fidelity device-scale simulations built using FLUENT computational fluid dynamics software. The advanced technology behind APECS

2.0 includes many new features that make it easier, faster, and cheaper for users to optimize existing and future plant designs with a high degree of confidence that predicted results will be realized. APECS 2.0 also allows the systematic generation of fast, accurate, CFD-based reduced-order models (ROMs), including time-averaged ROMs based on transient CFD simulations, using an innovative off-line ROM Builder. The seamless and tight integration of the ROM Builder with the recently released ANSYS® DesignXplorer™ provides even more new ROM features that make it easier, faster, and cheaper for engineers to optimize existing and future plant designs with a high degree of confidence. Developed in collaboration with ANSYS, Inc., the pioneering APECS 2.0 software won a 2011 R&D 100 Award and a 2011 Federal Laboratory Consortium Mid-Atlantic Region award for excellence in technology transfer. ALSTOM Power has been a valuable industrial partner and has applied APECS co-simulations to a wide variety of energy applications, including conventional pulverized coal combustion, oxy-combustion, IGCC, and chemical-looping combustion and gasification.

DID YOU KNOW

In the gasifier, high temperatures and pressures cause a series of chemical reactions between fuel (such as coal) and air and steam to produce synthesis gas (syngas). With further processing, syngas can be used to produce high-value fuels and chemicals. An important component of gasification is that CO_2 can be separated out of the resulting gas stream and be readied for utilization or storage.

Coal

NETL Invention Modifies Gasifier Refractory Microstructure—

Refractory service life is a major obstacle to wider use of gasification. NETL researchers have developed a way to modify commercial high-chrome oxide refractory material that could eliminate structural material separation (spalling)—a major cause of refractory wear and failure from slag penetration into the brick structure. In laboratory-scale rotary slag experiments, the NETL brick treatment limited slag penetration into the refractory microstructure to distances of 1–2 millimeters versus distances of up to 10–15 millimeters in untreated materials. Untreated refractory materials removed from commercial gasifiers have exhibited slag penetration depths over 60 millimeters, which can result in cracking, spalling, and loss of the penetrated area. NETL has an application pending with the U.S. Patent and Trademark Office for the brick structure modification, which is expected to significantly increase refractory service life.

Hydrogen Separation Membrane Projects Meet Phase-I Milestone—

Project teams at Worcester Polytechnic Institute of Worcester, MA, and United Technologies Research Center of East Hartford, CT, have each achieved a target production level of 2 pounds per day of hydrogen at 95 percent purity using proprietary laboratory-scale separation membranes operating in harsh coal-fired gasifier synthesis gas streams. For these and two other organizations that are developing hydrogen separation technologies in cooperation with NETL, the

Phase II goal is a minimum capability of 100 pounds per day of hydrogen to validate system design, production methods, and separation performance during long-term parametric testing at a gasification facility. It is anticipated that only the two top performing projects will advance to Phase II under a planned down-selection process scheduled in fiscal year 2012.

Refractory brick showing hexavalent chromium formation (yellow and green surface salt on brick).

NETL Thermodynamic Predictions Show Chromium Not an Issue in Reducing Environments—

Thermodynamic predictions indicate that formation of hexavalent chromium (Cr+6) will not be an issue in mixed carbon feedstock for gasification when coal, petcoke, or biomass are used. When potassium, sodium, and calcium levels in slag are typical of those found in mixed carbon feedstock, thermodynamic predictions indicate that chrome oxide in gasifier refractory brick would form high levels of Cr+6 with slag over temperatures ranging from room to gasification conditions in oxidizing environments similar to those found in cement or glass production. However, with the same slag chemistry and temperatures under reducing conditions such as found in gasification, Cr+6 levels were predicted to be well below levels of concern and, in most cases, below detection limits.

Novel Hydrogen Membrane Test Successfully Meets Target—

A novel hydrogen membrane material developed at Worcester Polytechnic Institute in cooperation with NETL has separated hydrogen with purity greater than 99.1 percent from a gasification stream at the National Carbon Capture Center. The gasification stream consisted of 48.4 percent nitrogen, 39.9 percent hydrogen, 6.6 percent CO_2, 3.8 percent carbon monoxide, 1.3 percent methane, and negligible hydrogen sulfide and carbonyl sulfide (dry basis). The single membrane tube successfully demonstrated a stable hydrogen flux in the range of 70.9–72.8 standard cubic feet per hour per square foot at 450 °C under a differential pressure of 150 pounds per square inch, which meets the critical Phase 1 hydrogen production target of 2 pounds per day.

Biomass Cogeneration Facility Installed—

Ameresco Federal Solutions, Inc., replaced four coal-fired, cold war-era boilers at the Department of Energy's Savannah River site with two massive boilers capable of burning wood chips and shredded tires. NETL helped negotiate the $795-million, 20-year contract that made the biomass cogeneration facility a reality—by far the biggest renewable energy contract signed by the federal government. The Energy Savings Performance Contract was made under the Federal Energy Management Program, which funds facility improvements with projected savings in energy costs. Ameresco expects the U.S. Government will save $944 million over the life of the contract. Designed to last more than 40 years, the new facility will provide roughly 30 percent of the power and all of the steam needed for heating and cleaning at the former nuclear bomb-production site. Energy savings beyond the 20-year agreement will redound to the U.S. Government.

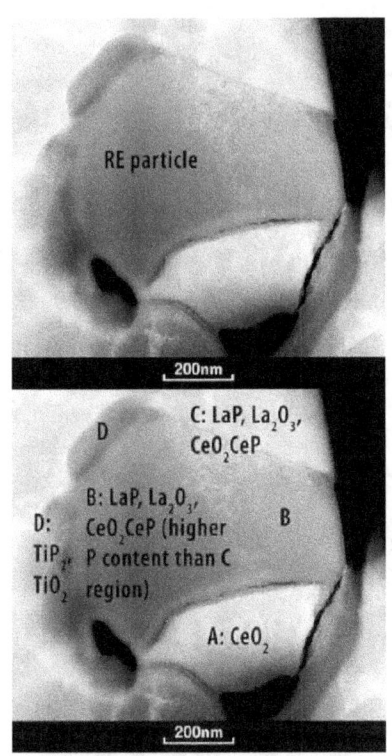

(top) Cross-sectional Scanning Transmission Electron Microscopy-High Angle Annular Dark Field (STEM-HAADF) image of an RE particle before oxidation. (bottom) The RE particle divided into 4 regions according to the chemical composition and the compounds present.

NETL Completes Study on Beneficial Effect of Rare Earth Additions to Iron-Chrome Alloys—Working in collaboration with NETL–Regional University Alliance researchers at Carnegie Mellon University, NETL scientists have developed a model to explain how adding a very small amount of a rare earth element (lanthanum, cerium, yttrium, etc.) to chromium oxide-forming iron- and nickel-based alloys can markedly improve oxidation resistance. In studying transient oxidation behavior of an iron, chromium, manganese, and rare earth alloy, the team found that the rare earth element diffused into the scale from rare

earth particles on the alloy surface during exposure to dry air at 800 °C. The effect is credited with significantly reducing scale growth rate and improving scale adhesion by causing an inversion of oxide growth inward under a coating of cerium oxide (CeO_2), reducing the outward chromium and manganese metal ion diffusion by trapping them in defect pairs with oxygen vacancies in the CeO_2 crystal structure. The study is detailed in 2011 in the peer-reviewed journal *Metallurgical and Materials Transactions* A (Vol. 42, No. 1). While the alloy composition selected was of interest as a metal interconnect for solid oxide fuel cells, iron-chrome alloys are also used extensively in fossil fuel power plants for boiler tubes, turbine rotors or discs, turbine blades, etc.

Advanced Research Materials Program Results in Approval of Inconel® 740—On the first ballot, the American Society of Mechanical Engineers (ASME) Boiler and Pressure Code Standards Committee voted to approve the use of precipitation-hardenable Inconel® 740 in wrought sheet, plate, rod, seamless pipe and tube, fittings and forgings in welded construction for service up to 800 °C under Section I rules. The nickel-based alloy has proven to be the best candidate material of construction for components such as steam headers and piping of a boiler operating at advanced ultra-supercritical steam conditions (760 °C and 5,000 pounds per square inch). Special Metals Corporation of Huntington, WV, developed the metal's optimum composition working with an NETL-supported collaboration led by Energy Industries of Ohio, with participation by major U.S. boiler manufacturers, the Electric Power Research Institute, and Oak Ridge National Laboratory to develop the materials technology necessary for construction and operation of advanced ultra-supercritical coal-fired boilers. Boilers

operating at these advanced conditions could reach efficiencies approaching 48 percent (up from the current domestic fleet average of 37 percent) and, if oxy-fired, would discharge sequestration-ready CO_2.

FYI

"Refractory" refers to heat-resistant materials, typically oxides, that have the ability to remain chemically and physically stable when exposed to high-temperature, severe environments. An example is the refractory liner used to protect gasifiers in integrated gasification combined cycle power plants. NETL is researching stable liner materials for the extreme service conditions of these and other future power generation systems.

Coal

Examining coal concepts at the atomic scale. NETL grew this 10 nanometer by 10 nanometer iron oxide particle to better understand the conversion of gasified coal into electric power and fuel. Individual atoms are seen as a pattern of small, gold-colored spots.

NETL Scientists Watch Atoms React at the Active Sites of Water-Gas Shift Catalyst

NETL researchers developed an in situ scanning tunneling microscopy method for visualizing with atomic-scale resolution a key reaction step occurring on an array of surface-grown iron oxide particles that simulated a real-world water-gas-shift (WGS) catalyst exhibiting a distribution of crystal sizes and structural defects. The technique manifested in nearly real time the incorporation of hydroxyl groups at particle edges as water molecules dissociated there. The oxygen of the hydroxyl group was observed to incorporate seamlessly into the iron oxide lattice with no significant change in atomic registry, illustrating that under-coordinated iron sites at crystal edges are responsible for the observed reactivity. Furthermore, hydroxyl groups were shown to deactivate these sites, suggesting that extremely water-rich atmospheres can kill the reactivity of the system studied. Described in *Langmuir's* March 2011 issue (Vol. 27, No. 6, pp. 2146–2149), the study's results provided direct experimental evidence that structural defects and under-coordinated sites at particle edges are indeed the active centers for WGS reactions. This new level of scientific insight will allow computational modelers and experimentalists to design and synthesize improved WGS catalysts with enhanced reactivity and selectivity.

NETL Study Examines the Physical and Policy Aspects of Frequency Instability in North American Electric Interconnections

A new report available from the NETL website correlates the increased number of larger and longer-lasting frequency excursions in North American interconnections with electricity market design and frequency control regulations. Even small deviations from 60 hertz can adversely affect sensitive end-use devices. While physical laws govern the frequency stability phenomenon and system control efforts are responsible for maintaining the nominal system frequency, the regulatory environment influences market design, which affects frequency stability, as well as policies that directly affect frequency control practices. The report covers both the technical and policy aspects of frequency stability.

NETL Completes Report on Role of Coal in a Smart Grid Environment

Smart Grid technologies will enable grid operators to manage supply and demand requirements, shifting load from peak-demand periods to periods of lower demand. This would allow expensive and inefficient peak generation to be replaced by less expensive, more efficient base load generation. NETL examined how that base load might change in a Smart Grid environment and the implications for coal in providing centralized generation,

distributed generation, and combined heat and power, including reserve and ancillary services. The study developed a model to demonstrate operational and economic characteristics of coal generation technologies in a "Smart Grid City of the Future." For one set of assumptions,

investing in a Smart Grid-enabled infrastructure by the model municipality would result in a payback period of approximately four years.

Long Island Power Authority Demonstration Evaluates Smart Grid Control Options—Long Island Power Authority (LIPA) is testing point-to-point (900 megahertz) and mesh (450 megahertz) networking technologies in the Bethpage and Hauppauge communities, respectively. The two tests have engaged 200 residential and commercial customers (100 per community). The communication networks include a front-end server at the vendor data centers, collectors or access points deployed on utility poles in local neighborhoods, and niche cards located in devices such as meters. Web-based energy usage information is supplementing the currently slower-performing mesh network for Hauppauge residential participants, but the point-to-point network has matured over the 18-month period and is now communicating efficiently. To date, 65 of 119 residences participating in smart "time of use" pricing have realized electricity savings of approximately $108 per household, with more savings expected through improved education and system maturation. One of the first projects to successfully demonstrate Smart Grid technology, this Congressionally directed project, funded by DOE'S Office of Electricity Delivery and Energy Reliability and managed by NETL, is demonstrating the hardware and software needed for a secure, interoperable, and open Smart Grid

while actually reducing energy costs for ratepayers.

Milestone Achieved in Integrating Distributed Resources with Electricity Delivery—Engineers at Consolidated Edison Company of New York successfully executed a planned test whereby a signal initiated at a demand response command center caused the transfer of electrical load at a remote facility from a commercial to an emergency bus bar. The transfer test was contingent on human authorization at the remote facility before automatic engagement of the equipment designated by the signal could proceed. The test demonstrated the utility's ability, when pre-arranged, to automatically control customer loads as a way of reducing peak power demand in emergency situations. The technology could allow a utility company to quickly and reliably activate customer-owned resources to respond to system emergencies, offering improved grid reliability, greater operational efficiency, and the option of deferring investments in distribution capacity. NETL manages this project on behalf of DOE'S Office of Electricity Delivery and Energy Reliability.

Water Recovery Technique Successfully Field-Tested— Slipstream tests of an innovative water-saving technology developed at Lehigh University in cooperation with NETL were completed successfully at three power plants—a wet scrubber-equipped, high-sulfur bituminous coal-fired plant and two unscrubbed, high-moisture, low-rank

coal-fired plants. The annual estimated return on investment for the condensing heat exchangers installed downstream of wet flue gas desulfurization (FGD) units is approximately 162 percent. Results

What is distributed generation?

"Distributed generation" refers to power produced close to where it is used, which reduces the amount of power lines required and the amount of energy lost in transmission. Distributed generation typically has less capacity than central station power plants, which historically have provided most of the electricity through major transmission and distribution systems.

suggest that, if installed upstream of wet FGDs or in plants with no wet scrubbers, condensing heat exchangers designed to cool flue gas to intermediate temperatures likewise have cost-savings potential. Whereas stack gases produced from coal firing are usually kept at approximately 300 °F to minimize acid corrosion and provide buoyancy, cooling them below the water vapor dew point would condense water for power plant use, recover latent and sensible heat, promote mercury removal, and reduce acid and water vapor content to mitigate the cost of capturing CO_2 in amine and ammonia CO_2 scrubbers.

Advanced Power Systems

Coal

Total magnetic field map of a coal waste impoundment in southern West Virginia.

Pioneering NETL Aerial Surveys Locate Potential Hazards at Coal Waste Impoundments—NETL

researchers and collaborators from the University of Pittsburgh have successfully applied multi-frequency helicopter-borne electromagnetic (HEM) surveys for the first time to determine the internal structure and integrity of mine-impoundment structures. Based on HEM surveys conducted at 14 coal-waste impoundments in West Virginia, the approach found the location and elevation of water-saturated zones more rapidly, economically, and thoroughly than the use of piezometers (instruments for measuring pressure), as currently specified by the Department of Labor's Mine Safety and Health Administration. Moreover, the HEM surveys identified areas of unconsolidated slurry buried deep beneath thick lifts of coarse coal waste—an unstable condition that can lead to rapid, catastrophic dam failure. Described in the

journal *Geophysics* on December 20, 2010, (Vol. 75, No. 6) published by the Society of Exploration Geophysicists, findings from the research suggest that HEM surveys could also locate failure-prone areas in dams surrounding coal ash impoundments.

Novel Approach Licensed for Relieving Congested Electric Transmission Lines—Smart Wire

Grid, Inc., has acquired the license to a new technology for increasing the electricity-carrying capacity of existing transmission lines, thus deferring the need to erect new lines. Developed with support from DOE's Office of Electricity Delivery and Energy Reliability and the Tennessee Valley Authority, the "smart wire" technology employs cylindrical modules that clamp around individual transmission lines and act inductively to resist the flow of current when the line's carrying capacity is approached, causing power to flow over underutilized lines in the grid. Each 5-foot-long module can change the impedance of a one-mile line segment by roughly 2 percent, and when distributed along the length of a transmission line, the modules can each be triggered at predefined set points to affect a gradual increase in line impedance. Depending on the system's level of sophistication, modules can be designed to operate autonomously or in concert. This distributed series reactance technology was developed by researchers at Georgia Institute of Technology for PSERC.

Methodology Increases Accuracy of Life Expectancy Estimates of Electricity Transmission and Distribution Transformers—In

support of PSERC, researchers at the

Colorado School of Mines have developed an optimization methodology to more accurately size and minimize the cost of liquid-filled transformers for new and retrofit transmission and distribution applications. Employing sensor and monitoring technology, and considering a unit's life, relative aging, and end-of-insulation-life criteria, the methodology calculates a transformer's hottest spot temperature, the top oil temperature, the bottom oil temperature, the loss-of-insulation life, the remaining life, and energy losses.

New Model Assesses Techno-Economic Impact of CO_2 Abatement Regulation—In support

of PSERC, researchers at Wichita State University and collaborators at the University of California, University College Dublin, and the California Independent

System Operator have developed a model that can forecast the cost impact and effectiveness of greenhouse gas regulation, considering regional sensitivity to carbon tax and use of renewable fuels. Preliminary results from using the model to simulate energy markets based on the Institute of Electrical and Electronics Engineers reliability test system show that modifying generating schedules (i.e., redispatch) in response to transmission line congestion would have a significant effect on CO_2 emissions. If the price of CO_2 were low, redispatch would involve units firing the same fuel, resulting in little effect on costs and CO_2 emission. On the other hand, a high CO_2 price will cause redispatch decisions to favor natural gas over coal, resulting in lower emissions but higher cost—perhaps three times greater than current electricity prices, not including the CO_2 tax payments. Individual utility regions with differing generation mixes and transmission constraints merit a focused analysis using the model.

Power Systems Engineering Research Center Provides Guidance for Electricity Transmission Modeling—As a

utility industry contributor to PSERC, Southern Company Services, Inc., has prepared a three-page, link-filled guide to help grid modelers find data and information more quickly for instructional or modeling purposes. Embraced by the PSERC Industry Review Board, the guide summarizes how to access power flow models and raw data for western and eastern interconnections, power management units, and supervisory control and data acquisition. The guide provides a succinct description of the data sources, advice on their use, and principles for maintaining data security and confidentiality.

Power Plant Cooling Technology Sells—SPX Cooling Technologies, Inc.,

has sold its first ClearSky™ cooling tower, developed under DOE's Innovations for Existing Plants–Water Management Research & Development program implemented by NETL. The hybrid air and water cooler performs like a conventional wet cooling tower, while avoiding the penalty associated with the higher condenser discharge temperature and turbine back-pressure typical with dry cooling systems. The ClearSky product, which emerged from the Air2Air™ prototype, is initially aimed at the plume abatement market. Many localities require plume abatement from cooling towers to mitigate unsightly and potentially dangerous visibility and icing conditions that can develop as water vapor from the plume condenses. Depending on climatic conditions, ClearSky technology can recover 15–25 percent of the water that would otherwise be lost from an evaporative cooling tower, with less energy than conventional coil-type plume abatement products. A typical 300-megawatt coal-fired plant with evaporative cooling towers could save an estimated 600,000 gallons per day of freshwater by using the technology. If all industrial cooling towers in California were retrofitted with ClearSky technology, roughly 188 million gallons of high-quality freshwater could be conserved daily—increasing residential water capacity by 7.6 percent.

DID YOU KNOW

Many aspects of our nation's aging power grid need to be updated and modernized. Smart Grid can upgrade the electric power system and provide reliable, safe, and economical power by integrating sensors, communications, and controls that effectively manage emerging grid assets, such as renewable energy, electric vehicles, energy storage, distributed generation, and smart appliances.

Advanced Power Systems

Turbines

Advanced turbines are critical components to enabling U.S. electric power production that is efficient, affordable, secure, and adaptable to CO_2 capture. NETL's turbine research is innovating technologies for fuel-flexible power generation, for reducing NO_x emissions, and for improving performance.

CES's oxy-fuel gas turbine, the OFT-900 is a key component of CES's oxy-fuel power generation system, which is capable of capturing greater than 99 percent of CO_2 produced in the combustion of a variety of fuels, including coal syngas and natural gas.(Image courtesy of Florida Turbine Technologies, Inc., and Clean Energy Systems, Inc.)

Test Site for the Clean Energy Systems Advanced Turbo Machinery Project Selected Ahead of Schedule—Management at Clean Energy Systems (CES) of Rancho Cordova, CA, has selected the CES Kimberlina Power Plant near Bakersfield, CA, for testing the OFT-900 oxy-fuel turbine they are developing with Recovery Act funds as part of the NETL Industrial Carbon Capture and Sequestration portfolio. The CES power generation concept utilizes a proprietary oxy-fuel combustor derived from rocket engine technology that produces a high-temperature and high-pressure working fluid of steam and CO_2 from the combustion of coal synthesis gas and other fossil fuels; this fluid expands in a series of power-generating turbines—with the OFT-900 envisioned as the intermediate-pressure turbine in the system. Since the working fluid is undiluted with nitrogen, carbon capture for utilization or storage is relatively simple. In addition, the approach is attracting commercial interest for near-term applications in enhanced oil recovery. Copenhagen-based global oil and gas producer, Maersk Oil, licensed the CES oxy-fuel combustor technology in January, 2011, and they announced an agreement with Siemens Energy, Inc., to develop oxy-fuel gas turbines for Maersk's

"TriGen" polygeneration technology. In the TriGen application, oxygen provided by an air separation unit combusts with contaminated, low heating-value natural gas (such as that found in stranded gas fields) to produce the drive gas for the Siemens turbine, which in turn produces electrical power and an exhaust easily treatable to yield CO_2 for near- to mid-term commercial applications such as enhanced oil recovery and enhanced gas recovery. Maersk asserts that using CES technology with an intermediate-pressure oxy-fuel turbine of the scale being designed, built, and tested in this project will be more economically viable than the alternative CO_2 stripping practice for stranded gas fields, particularly those offshore. Earlier-than-planned acquisition of a used Siemens SGT-900 gas turbine to be converted for oxy-fuel operation, selection of the test site, and the Kimberlina Power Plant permit finalization may enable OFT-900 turbine testing to begin up to 1 year sooner than projected.

Low-Swirl Technology Demonstrates Fuel Flexible Operation—A fully functional prototype of a simple advanced-combustion concept from Lawrence Berkeley National Laboratory that is being developed for the gas turbines

in an integrated gasification combined cycle clean-coal power plant was evaluated operationally in terms of fuel flexibility, ease of light-off, loading, combustion dynamics, and pollutant emissions. Test results indicated that the low-swirl injector design—incorporating proprietary premixer technology with three fuel circuits and Lawrence Berkeley's flared nozzle—could support stable operation with both natural gas and a blend of 65 percent hydrogen and 35 percent nitrogen at pressures up to 20 atmospheres, while maintaining low-combustion dynamics and NO_x emissions well below 10 parts per million (corrected to 15 percent oxygen). The sub-scale, low-swirl prototype was developed with collaborators at United Technologies Research Center and Pratt & Whitney Power Systems under an NETL-supported field work proposal.

Optimized Nozzle Geometry Lowers NO_x Significantly in Turbine Test—As part of the Office of Fossil Energy's Advanced Integrated Gasification Combined Cycle (IGCC)/Hydrogen Development program implemented by NETL, GE Energy researchers demonstrated another significant reduction in NO_x emissions during a full-can rig test conducted at advanced gas turbine operating conditions

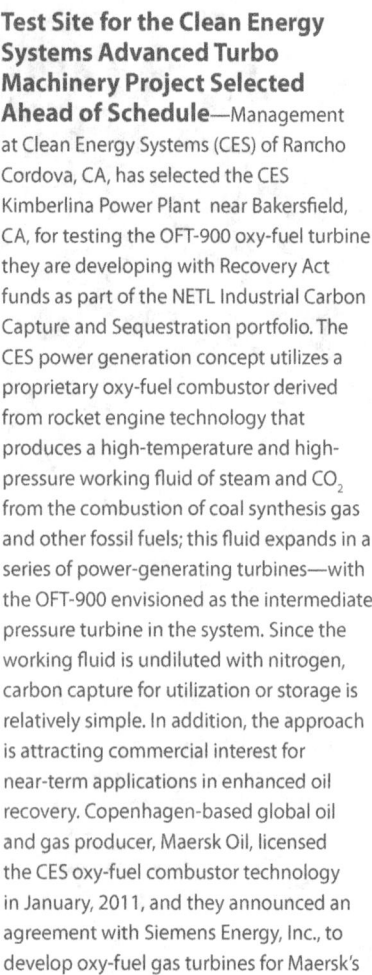

with high-hydrogen fuel. Compared to past performance, a reduction of nearly 25 percent in corrected NO_x emissions was credited to optimized fuel nozzle shape and sizing that eliminated cooling and leakage air, which in previous designs had bypassed the nozzle. Over the past 18 months, the project team has lowered NO_x levels from parts-per-million in the low double-digit range to a level approaching the 2 parts per million program target, when available nitrogen diluent is utilized. At the same time, operability in bench tests with high-hydrogen fuels has improved. Reliable operation without flashback has been demonstrated with 100 percent hydrogen fuel diluted with nitrogen, and the test rig now has been fired for a total of more than 80 hours with greater than 85 percent hydrogen by volume in the fuel reactants. These advancements represent significant progress toward eliminating the need for after-treatment, saving roughly $4 million in capital cost per gas turbine.

CMC shrouds in first stage of a GE Energy F-class utility gas turbine at a JEA utility in North Florida.

New Material Improves GE Turbine Products

A 2011 report shows that the 2006 installation of a full set of silicon carbide-based shrouds in the first stage of a GE Energy F-class gas turbine resulted in significant performance improvement at a Jacksonville Electric Authority utility in north Florida. Invented at the GE Global Research Center in Niskayuna, NY, and developed as part of an NETL-managed project funded by the Office of Electricity Delivery and Energy Reliability, the melt-infiltrated ceramic matrix composite (CMC) permits higher

operating temperature compared to that of metal components. Higher operating temperature enables greater gas turbine power output and efficiency with lower emissions and longer intervals between scheduled maintenance. GE is incorporating CMC components into its product line of aircraft and utility gas turbines.

Micro-mixing Fuel Injector System Finalized for Ultralow NO_x Emissions

Based on an earlier 1-megawatt (thermal) fuel injector design that showed excellent emissions performance, researchers at Parker Hannifin Corporation working in cooperation with NETL have designed a modular, two-stage micro-mixing fuel injector for combustion turbines burning high-hydrogen fuel. The improved design enhances operability by incorporating fuel staging, where the primary fuel circuit is recessed and oriented relative to the secondary fuel circuit so that the primary flow is sheltered from the secondary flow. The final design configuration also allows for operation with a high-bandwidth "piezovalve"—already under development at Parker—that modulates fuel-mass flow to extend the fuel's lean stability limit while controlling combustion dynamics. Atmospheric testing at the Georgia Institute of Technology with the piezovalve modified for high-hydrogen fuels and integrated with the improved injector confirmed the benefits of the recessed primary fuel circuit and the system's ability to control low-frequency combustion dynamics.

Gas Turbine Test Demonstrates Low NO_x Emissions with Hydrogen Fuel

As part of an NETL-managed project, a full-scale combustion test of rich catalytic lean-burn (RCL®) technology produced low single-digit NO_x emissions in the adiabatic flame temperature range 2,200–2,750 °F. The test was conducted at Solar Turbines

Incorporated using a high-pressure blend of hydrogen and nitrogen. Developed at Precision Combustion, Inc., North Haven, CT, RCL technology features catalytic pilots to produce leaner, stable overall combustion and very low NO_x emissions without postcombustion controls or efficiency penalty. The successful demonstration places the technology among the options for pursuing the 2 parts-per-million NO_x target of DOE's Hydrogen Turbine Program.

DID YOU KNOW

Turbines have been the world's energy workhorses for generations, harkening back to primitive devices such as waterwheels (2,000 years ago) and windmills (over 1,000 years old). Today, turbines not only power aircraft and vehicles of all sorts, they are the heart of almost all of the world's electric generating systems. Steam and gas turbines are mainstays of large-scale electrical power generation. In pursuit of higher efficiency, superior environmental performance, and lower cost, NETL's in-house turbine research and development group uses unique facilities to evaluate new concepts in combustion and turbine materials.

Turbines

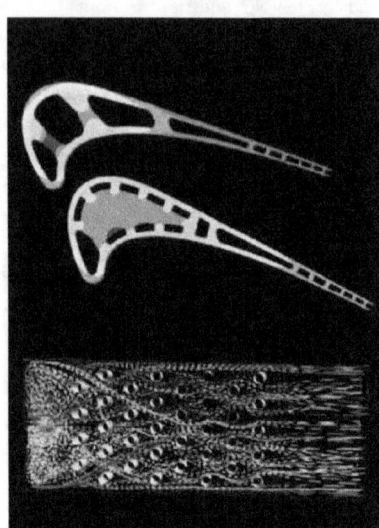

The development of three-dimensional computational models of heat transfer gives vital insight on turbine airfoil life to researchers.

Study Focuses on Turbine Airfoil Life—Researchers at NETL, along with University of Pittsburgh collaborators in the NETL-Regional University Alliance (NETL-RUA), have studied the effects of compressive creep strain on the performance of turbine airfoil superalloys and thermal barrier coatings at elevated temperatures. The investigation complements other airfoil studies that typically examine either tensile creep or thermal cycling behavior. The team conducted laboratory-scale experiments in conjunction with computational model development to verify the underlying damage mechanics concepts using nickel-based single crystal René N5 coupons with a commercial bond coat and thermal

barrier coating systems. Oxidation tests were conducted at temperatures ranging from 900 °C to 1,100 °C for periods of 100 to 3,000 hours in slotted silicon carbide fixtures. The difference in the coefficients of thermal expansion of the René N5 substrate and the test fixture was found to introduce heat-induced compressive stress in the coupon samples. Results of the investigation appear in the April 2011 issue of *Journal of Engineering for Gas Turbines and Power* (Vol. 133, No. 9)

Second-Generation Water-Guided Laser Drilling System Becomes Operational—In an NETL-managed project, Physical Sciences, Inc., (PSI) completed construction of a second-generation water-guided laser system for high-speed drilling of cooling holes in critical gas turbine parts. The key element of the system is the water-guiding component that creates a high-pressure stream of water to guide a high-power laser on a work surface. PSI used computational fluid dynamics to design the component for operation, with water pressures up to 3,000 pounds per square inch—without causing bubble-forming cavitations, which would disrupt the light-guiding capability of the water stream. The system demonstrated that it can drill cooling holes through Inconel 718 (a hardened austenitic nickel chromium-based superalloy) in less than 3 seconds, which is consistent with rates required for the production of gas turbine components. Hole drilling with continuous wave operation of the laser at 2,000 watts, as well as operation with 10 millisecond pulses at 1,000 watts, was demonstrated.

NETL researcher Dr. Michael Buric and Dr. Steven Woodruff assembling the gas composition sensor.

Unique Raman Gas Analyzer Made Portable—NETL innovators have constructed the first portable, field-ruggedized version of their Raman gas composition monitor developed for rapid, online fuel gas analysis. Real-time gas composition analysis will allow control systems for power equipment, such as gas turbines and fuel cells, to quickly and intelligently react to fuel composition changes for optimal performance with fuels, including coal-derived synthesis gas, coal-bed methane, biogas, and natural gas. The Raman gas composition monitor features metal-lined capillary sampling tubes that act like mirrored waveguides to greatly enhance weak Raman light-scattering signals for near-instantaneous analysis of gaseous mixtures with high accuracy.

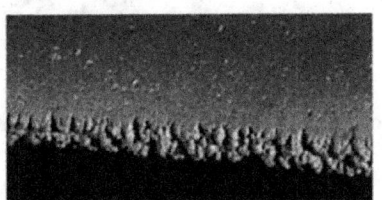

Leading edge cacitation exhibited on 1" stage fan blades resulting in poor aerodynamics and fuel consumption.

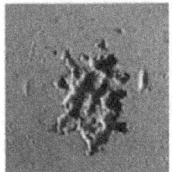

Simulated fluid erosion test

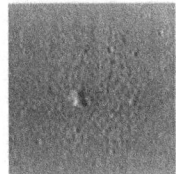

With nanocoat protection

(Images courtesy of MDS Coating Technologies Corporation.)

New Low-Cost Nanocoating Saves Fuel, Lowers Turbine Maintenance Cost—Product

developers at MDS Coating Technologies Corporation, working in cooperation with NETL materials scientists, have completed key erosion, corrosion, and fatigue testing of a nanostructured material that, when layered onto airfoils by a cathodic arc physical vapor deposition process, improves turbine operational efficiency and extends service life. The erosion-resistant nanocoatings are so thin that they impact neither the dimensions nor weight of the coated components, which is particularly important in compressor section gas turbine applications. The NETL–MDS collaboration was supported under the Nanomanufacturing Initiative of DOE's

Office of Energy Efficiency and Renewable Energy. Successful deployment of this technology will result in major petroleum savings for the U.S. commercial air fleet and enable the full energy efficiency benefits of employing inlet fogging, which can augment power output for an electricity generating plant by 7–8 percent under certain conditions. Moreover, a 1 percent increase in compressor efficiency decreases CO_2 emissions by 1–4 pounds of mass per hour.

Collaboration Develops Extreme-Temperature Thermal Barrier Coating—NETL collaborators at NASA

recently completed the first qualification testing of an extreme-temperature composite architecture for advanced hydrogen-fired turbine applications. The composite consists of a nickel-based single crystal René N5 substrate material, NETL's Coatings for Industry A1D bond coat system, a commercially applied state-of-the-art air plasma sprayed yttria-stabilized zirconia layer, and NASA's extreme temperature overlayer. In the NASA laser thermal flux tests, the external surface temperature of the overlayer reached approximately 1,480 °C, while the bond coat interface temperature was maintained at around 1,120 °C by back-side cooling of the test coupon. The coating remained adherent and intact even after 50 cycles of heating for a period of one hour followed by cooling to room temperature in 3 minutes.

What are superalloys?

Superalloys are metal alloys that have been developed to withstand intense heat and/or pressure in applications, such as turbine blades, for advanced power plants, aerospace, and jet engines.

Fuel Cells

Suitable for a variety of power generation applications, solid oxide fuel cells can ensure and enhance our continued use of domestic coal—a primary resource for reducing U.S. dependence on imported oil—in a more environmentally friendly way.

NETL's "Hyper" facility assesses the performance of fuel cell-turbine combined systems.

Coal-Based System Designs Meet Cost and Performance Targets for Integrated Gasification Fuel Cell Power Plants

—The baseline integrated gasification fuel cell (IGFC) power plant conceptualized by SECA industry team leader FuelCell Energy will generate a nominal 670 megawatts of power from the coal synthesis gas fuel produced by catalytic gasification and warm gas cleanup processes. The power island combines a solid oxide fuel cell (SOFC) with a steam bottoming cycle. CO_2 produced from oxy-combustion of the SOFC anode exhaust will be readily captured. The optimum electrical efficiency of the system is 58.7 percent (based on high heating value of coal) while capturing virtually all of the synthesis gas carbon as CO_2. The baseline plant consumes 75 percent less water than an equivalent pulverized-coal combustion plant, and it has an installation footprint comparable to that of an integrated gasification combined cycle plant. FuelCell Energy projects an SOFC power island cost of $635 per kilowatt (2007 dollars) and a stack cost of $147 per kilowatt, beating the SECA cost targets for central power generation. This breakthrough SOFC stack factory cost estimate results from a new generation of fuel cells developed by FuelCell Energy's partner, Versa Power Systems, Inc., based on improved materials and thin cell technology. Compared to earlier generations, the new generation of cells has demonstrated higher power density, an expanded range of operating temperature, and improved long-term stability with reduced power degradation. Advances in fuel cell manufacturing, including automation, process modification, and the better material utilization that results from higher cell production yield, also contribute to lower SOFC stack cost.

NETL-Led Alliance Develops Efficient Integrated Gasification Fuel Cell Technologies

—The SECA industry team led by UTC Power developed three integrated gasification fuel cell designs that produce greater than 100 megawatts net A/C power with efficiency greater than 50 percent and with carbon capture greater than 90 percent. Based on a preliminary analysis, the power block hardware for the baseline plant is expected to have a capital cost of $685 per kilowatt (2007 dollars), which meets the SECA cost target of $700 per kilowatt. The power block is assembled from "Gen 4" stacks developed by Delphi Automotive Systems. The new 40-cell Gen 4 stack demonstrated a maximum initial power of 6.4 kilowatts at an average power density of 398 milliwatts per square centimeter and average cell voltage of 0.7 volts, utilizing the SECA-simulated coal gas blend. Delphi has developed low-cost, high-volume manufacturing processes for Gen 4 stack components and has fabricated and tested multiple Gen 4 stacks in a variety of test

conditions. In completing a durability test lasting more than 2,500 hours, the Gen 4 stack demonstrated a degradation rate of about 4 percent in the first 300 hours and no measurable power degradation thereafter. The stack has also successfully completed 60 deep thermal cycles—from a 750 °C operating temperature to less than 100 °C—with a total power degradation of less than 5 percent.

NETL Identifies Options for Robust Control of Fuel Cell-Turbine Hybrids—Scientists operating the NETL Hybrid Performance facility compared multiple strategies for controlling the integrated operation of a fuel cell and turbo-machinery. The facility is designed for hardware-in-the-loop simulation of fuel cell dynamics integrated with turbine and compressor hardware. Transient experiments characterized dynamic system response with four decentralized control strategies of varying complexity. These were compared with two centralized (multiple input, multiple output) control strategies. While none of the decentralized control methods could achieve acceptable performance and disturbance rejection, a centralized approach with anti-windup and feed-forward compensation was successful. Described in the American Society of Mechanical Engineers July 2011 issue of *Journal of Engineering for Gas Turbines and Power* (Vol. 133, No. 7,), the investigation provides a key advancement in implementing a hybrid approach for ultra-efficient power generation.

Solid Oxide Fuel Cell Research Generates Commercial Interest—A novel electrodeposition process being developed under an NETL-managed Small Business Innovation Research (SBIR) grant for coating SOFC interconnects has generated interest among several commercial SOFC developers, including FuelCell Energy,

Siemens Power Generation (for non-fuel cell applications), and GE Global Research. This R&D 100 Award-winning technology could enable the use of low-cost ferritic stainless steels in planar SOFC stack construction. The faradaic process would increase SOFC system longevity and efficiency by reducing cathode degradation while optimizing component machinability, system oxidation resistance, electrical conductivity, and coating thickness. Data show that a test cell with interconnects coated with a manganese-cobalt spinel material using this environmentally friendly process exhibited a lower degradation rate relative to uncoated interconnects. Moreover, electroplating is easier to employ and lower in cost compared to other coating methods. The process emerged from research by Faraday Technology, Inc., with NETL and West Virginia University.

Novel Solid Oxide Fuel Cell Interconnect Coating Material Achieves Performance Milestone—A manganese-cobalt spinel interconnect coating material for low-cost ferritic stainless steel (AISI 441) has reached the two-year mark in a small-scale, long-term test at 800 °C. Electrical resistance while in contact with a representative cathode contact material has remained low and stable throughout the test, which is continuing toward a 40,000-hour target—the projected operational lifetime of a SOFC stack. The coating material is being evaluated as

part of an NETL-supported Field Work Proposal at Pacific Northwest National Laboratory in support of the SECA Core Technology program. The cost of large-scale manufacturing and processing of AISI 441 is expected to be an order of magnitude less than for state-of-the-art high-temperature metal alloys suitable for the severe environment of SOFC interconnect service.

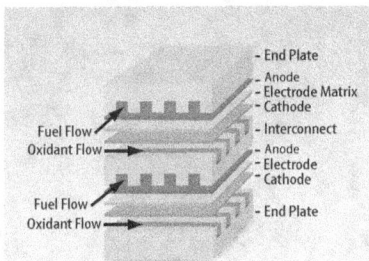

What is the future of SOFCs?

Solid oxide fuel cells (SOFCs) are being developed that will produce clean, efficient, low-cost power for large stationary applications, including central generation. In support of this goal, NETL and partners are working to improve the endurance and reliability of SOFCs to commercially viable levels.

Clean Energy

Sorbent pellets of immobilized polyethylenimine on the support CARiACT Q10, a commercial silica gel support with a diameter of 100 to 350 μm. Sorbent pellets were prepared in two 600-lb batches by Pressure Chemical Company.

Reducing emissions and conserving energy, while building a portfolio for sustainable energy production.

Simply put, reducing emissions and byproducts during energy production results in a cleaner environment. A complementary approach is managing demand-side efficiency through technologies that reduce energy consumption. While alternative energy sources generate emission-free power, these resources cannot yet fulfill America's energy demands. Fossil resources, especially coal, are necessary for the foreseeable future to power the United States and developing countries. Therefore, a major component of NETL's mission is to reduce the environmental impact of fossil resources. NETL has long been involved in research, development, and deployment of technologies that make energy production more environmentally friendly. Through basic research and partnership with others, NETL researchers are developing improved, economic, and effective methods of carbon capture using solvents, sorbents, membranes, and other technologies with the goal of creating cost-effective systems capable of capturing 90 percent of CO_2 from stationary sources by 2020.

ARRA-Funded Research Ignites Economic Growth

Novel Energy Storage Entrepreneurship

In 2007, Dr. Whitacre of Carnegie Mellon University had an idea to research low-cost electrochemical couples. Five years later that idea turned into Aquion Energy. The name is a combination of "aqueous ion," and represents the company's hard work and successes in creating the sodium-ion battery. (Image courtesy of Aqyion)

NETL is advancing energy options to fuel the U.S. economy, strengthen national energy security, and improve the environment. In 2009, the American Recovery and Reinvestment Act (ARRA) was set in motion with goals to create jobs, spur the economy, and foster government spending, and NETL was an ideal candidate for funding. With capabilities to address national energy challenges and strong ties to regional energy partners, the Lab is managing multiple projects that are turning this funding into successful energy solutions.

One NETL ARRA project in particular captured industry's attention and earned the Pittsburgh Technology Council's 2011 Tech 50 Award for Start-Up of the Year. The award honors companies that demonstrate strong growth and advancement in innovative products or technologies. Winning this award was Aquion Energy's sodium-ion battery: a project managed by NETL in support of the research and development program for DOE's Office of Electricity Delivery and Energy Reliability. NETL is a well-established leader in energy research, and the Lab is pleased that its management role could help a local startup gain national recognition and vitalize the Pittsburgh region as a go-to for energy solutions.

Aquion's environmentally friendly sodium-ion battery with novel ambient-temperature technology is a revolutionary achievement. The battery supplies and stores energy using sodium, which is a safer, more abundant, and cheaper alternative to the more commonly used lithium. Powering our everyday devices, such as mp3 players, cell phones, and laptops, lithium batteries perform well, but they are too expensive for large-scale use. Aquion's sodium-based battery can store wind and solar energy at a low cost and in large amounts. During peak demands, the stored energy can be used to power our electrical grid. In addition to these benefits, the sodium-ion battery project is also expected to create more than 500 jobs by 2014 and 1,000 jobs by 2017. Not bad for a young company from Pittsburgh.

With an accomplished team of project managers, NETL was able to provide valuable coaching to a number of additional ARRA-funded projects focusing on carbon capture and re-use, innovative gas turbine manufacturing, novel turbine development, Smart Grid technology, and solar energy technology. During the timeline of these projects, NETL managed and sustained partnerships with private industry and well-known universities from across the nation, helping them to—

- Analyze results and develop detailed designs for pilot plants

- Install test wells and collect geologic and geophysical data from the wells

- Quantify formations for commercial carbon storage and possible enhanced oil and gas recovery

- Initiate construction activities aimed at demonstrating CO_2 capture processes for storage

- Launch multiple Smart Grid projects

In 2009, the Lab was entrusted with ARRA funding and charged with returning that investment to our nation by guiding innovative technologies toward commercialization; Aquion's sodium-ion battery is just one success. The ongoing, productive collaboration between NETL and its partners will continue to attract generous funding that yields novel technologies with benefits like job creation, energy reliability, and cost reductions. By recognizing the strengths of its energy partners both local and national, NETL has helped cultivate a winning region and also fortified the future of energy business for the nation.

Background image: Breadbox-sized sodium-ion batteries weigh approximately 50 pounds each and can be grouped together to power large electrical grids. (Image courtesy of Aquion)

Carbon Capture, Utilization, and Storage

Captured CO_2 can be compressed and either utilized, such as during enhanced oil recovery, or directed into underground porous rock structures for permanent secure storage. Both options keep the greenhouse gas out of our atmosphere, enabling a cleaner use of coal.

NETL Analysis Shows High Benefit for U.S. Economy and Environment with CO_2 for Enhanced Oil Recovery

—An analysis completed by collaborators at NETL and Advanced Resources International, Inc., revises a previous national resource assessment for using CO_2 for enhanced oil recovery (CO_2 EOR), employing an industry-reviewed methodology and information on the latest technological developments. Using current technology, 1,200 reservoirs in the contiguous states—not counting offshore fields, Alaska, and residual oil zones—are amenable to CO_2 EOR. Assuming a crude oil market price of $85 per barrel, these reservoirs represent 24 billion barrels of economic reserves. Under a next-generation scenario—with directional drilling and sophisticated, real-time diagnostics to inject large amounts of CO_2 at high rates—the economic supply from CO_2 EOR increases to 60 billion barrels. The revised assessment implies a strong dependence on carbon capture technology to supply the EOR floods, which could require 90 percent CO_2 capture at coal-fired power plants with an aggregate capacity of 50–80 gigawatts. Often viewed inaccurately as a niche opportunity, CO_2 EOR has the potential to be a strategically significant source of energy security for the United States as well as a platform for case studies and learning in CO_2 capture.

NETL Process Utilizes Captured CO_2

—Researchers at NETL have incorporated CO_2 into polymeric materials similar to epoxy. Like epoxy, these materials are made by mixing two liquid components: a multifunctional cyclic carbonate and a multifunctional amine. Using novel chemistry, the CO_2 is incorporated into the polymeric materials through the multifunctional cyclic carbonates. The two liquid components react to form a cross-linked polymer (hydroxyurethane) that is strong and adheres well to aluminum and glass surfaces. These materials also have potential application as sealants and binders for composites such as fiberglass and plywood. By developing technology that utilizes CO_2, NETL is working to reduce the quantity of the greenhouse gas that must be geologically stored and provide industry a means to offset storage costs partially through product sales

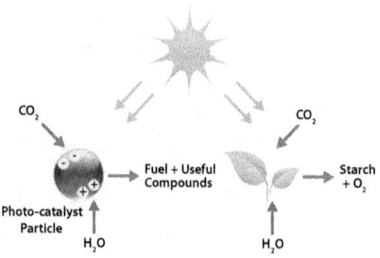

A semiconductor such as TiO_2 is used as a manmade photocatalyst to convert CO_2 into useful materials. The process is similar to how natural plant chlorophyll converts CO_2 into starch and O_2.

NETL Makes Breakthrough for CO_2 Reuse

—Scientists at NETL devised a successful strategy for sensitizing titanium dioxide (TiO_2) catalysts with efficient light absorbers that may make photocatalytic reduction of CO_2 practical under visible light irradiation. The team used lead sulfide (PbS) quantum dots (QDs) to enable photocatalytic conversion of CO_2 at frequencies ranging from the violet to the orange-red edge of the electromagnetic spectrum, depending on quantum dot size. This breakthrough represents the first demonstration of converting CO_2 at such low energy (orange light irradiation) levels using solid state photocatalysts. Under a broad band of illumination, the PbS quantum dots enhance CO_2 photo-reduction rates with TiO_2 by a factor of about 5 compared to unsensitized photocatalysts. Using the visible light (low-energy) portion of the electromagnetic spectrum for converting CO_2 and water into methane, methanol, and other value-added chemicals is a promising carbon utilization approach that can generate revenue to offset the cost of implementing CO_2 capture technologies. This significant advance in the development of new technologies for CO_2 utilization is detailed in the September 2011 issue of *Journal of Materials Chemistry* (Vol. 21, No. 35).

NETL Study Analyzes Uncertainties Associated with Greenhouse Gas Emissions Estimates

—The Energy Independence and Security Act of 2007 requires that any alternative fuel procured by the federal government be certified as producing fewer emissions than conventional petroleum-based fuels. Because greenhouse gas (GHG) estimates are often reported as point values for specific scenarios, the range of possible outcomes for a given fuel substitution is largely unknown. Now, NETL has developed a framework for incorporating uncertainty analysis when estimating the life cycle GHG emissions from the production of biomass. The approach will enable improved investment decisions that are based on estimates of net GHG reductions expected from substituting bio-based fuels for fossil fuels. The study is described in the December 2011 issue of the peer-reviewed Elsevier Publication, *Biomass & Bioenergy* (Vol. 35, No. 7).

Schematic of an ionic liquid interaction with CO_2.

NETL Computational Chemistry Group Expands Screening Methodology to New Classes of Solid Sorbents for CO_2 Capture—

NETL researchers have developed a novel computational technique for screening solid sorbents for CO_2 capture applications. This method describes the thermodynamic properties of different CO_2 capture reactions at various temperature and pressure conditions. Described earlier (August 2010 issue of *Journal of Chemical Physics*, Vol. 133, No. 7. and the January 2009 issue of *Physical Review B*, Vol. 79, No. 1), this technique has the advantage of identifying thermodynamic properties of the CO_2 capture reactions without any experimental input beyond crystallographic structural information of the solid phases involved. This computational method has been used to describe the CO_2 adsorption properties of different alkali-metal oxides, hydroxides, zirconates, silicates, and carbonates or bicarbonates and agrees well with experimental findings. Detailed results are described in the February 2011 issue of Elsevier's *Journal of Solid State Chemistry* (Vol. 184, No. 2), the September 2011 issue of the American Physical Society's journal, Physical Review B (Vol. 84, No.10), and the January 2011 issue of the American Institute of Physics's *Journal of Renewable and Sustainable Energy* (Vol. 3, No. 1).

NETL Provides Atomic-Scale Evidence of CO_2 Interaction with Photocatalyst—

Employing an improved version of density functional theory complemented by scanning tunneling microscopy measurements, scientists at NETL and University of Pittsburgh completed a comprehensive analysis of the adsorption properties of CO_2 on the surface of titanium dioxide as a function of coverage, surface sites, and the size of different slab models. This study provides essential information about various adsorption sites and the role played by surface oxygen defects and, for the first time, brought calculated binding energies into quantitative agreement with experimental results. The breakthrough is credited with correctly describing dispersion interactions between CO_2 and the oxide surface. The atomic scale theoretical predictions obtained were further confirmed by low-temperature scanning tunneling microscopy, which provided direct evidence of the atomic surface binding sites and confirmed the enhanced molecular binding of CO_2 at surface defect sites.

NETL Study Provides Insight for Optimizing Activation of Catalysts—

NETL researchers have used first principles density functional theory calculations to provide direct insight into how alkali promoters affect carbon monoxide activation properties of iron-based Fischer Tropsch catalysts. Systematic analysis of the molecular-induced stabilization effects, molecular vibrational shifts, diffusion properties, and modifications of activation barriers have been obtained as functions of different coverage conditions of the promoter species. Described in the February 2011 issue of Elsevier's premier journal for surface-related properties, *Surface Science* (Vol. 605, No. 3-4), the study identified the optimal conditions for the reaction, which could provide significant economic benefits by decreasing carbon monoxide activation energies as much as 31 percent when converting synthesis gas to hydrocarbons.

10-20%

What is enhanced oil recovery?

Enhanced oil recovery (EOR) is a set of technologies for increasing the amount of crude oil recovered from a reservoir. By gas (typically CO_2), chemical, steam, or microbial injection, an additional 10–20 percent of the remaining oil can be produced. One NETL-managed EOR project increased production from 16 barrels to more than 65 barrels of oil per day in the 15-acre project area.

Carbon Capture, Utilization, and Storage

Office of Fossil Energy Releases Updated Carbon Sequestration Atlas of the United States and Canada

—The new Carbon Sequestration Atlas of the United States and Canada (Atlas III) presents updated information on the location of stationary CO_2 emission sources as well as the locations and storage potential of various geologic formations across the majority of the U.S. and portions of Canada, as evaluated by the seven NETL-managed Regional Carbon Sequestration Partnerships (RCSPs). Estimates suggest that the formations discussed could store from 535 to 5,900 years of annual CO_2 emissions from the sources identified. The updated Atlas III outlines the DOE Carbon Sequestration program and international carbon capture and storage (CCS) collaborations, worldwide CCS projects, and CCS regulatory issues. Available online, Atlas III also provides information on the National Carbon Sequestration Database and Geographic Information System and commercialization opportunities for pertinent technologies within each of the RCSPs. Produced through the cooperation and coordination of carbon storage experts from local, state, and federal agencies, as well as industry and academia, Atlas III documents the great promise of carbon capture and storage technologies for addressing climate change while meeting the energy demands of an ever-increasing global population.

Low-Cost Methodology Could Assess Fate of CO_2 Used in Carbon Capture, Utilization, and Storage

—Working in cooperation with NETL, a research team from the University of Miami has deployed solar-powered global positioning system (GPS) stations on the ground's surface and seismic detectors in shallow underground vaults to monitor CO_2 injection operations at the Hastings West enhanced oil recovery (EOR) site. Approximately 30 miles south of Houston, TX, the site is located where Denbury Resources Inc. (DRI) plans to begin oil extraction in 2012. The team will also integrate surface geochemical sensor data and interferometric synthethic aperture radar satellite (InSAR) imagery with the GPS geodetic and seismic data recorded over several years into a straightforward series of procedures and algorithms to assess the cost and efficacy of tracking the CO_2 with this innovative approach. DRI operators are placing a new $1 billion pipeline in service to transport natural CO_2 from the Jackson Dome in Mississippi to Hastings for injection to depths of about 5,000 to 6,000 feet. The integrated monitoring system has the potential to detect real-time changes—such as ground displacements and seismic velocity anomalies in the reservoir and overburden—attributable to the injection of commercial quantities of CO_2.

Key Facility Commissioned for Large-Scale Midwest CO_2 Storage Field Test

—The facility that will capture, dehydrate, and compress a portion of the CO_2 stream from the ethanol production plant at the Archer Daniels Midland (ADM) corn processing complex in Decatur, Illinois, is fully commissioned and feeding the CO_2 to an injection well located on ADM property. The new facility dries the 99 percent pure CO_2 from the ethanol fermenters and compresses it to a supercritical dense fluid. The compressed CO_2 is transported to an injection wellhead through a 6,000-foot, 5-inch diameter carbon steel pipeline. In early November, the Illinois Environmental Protection Agency, which regulates ADM's Class I Underground Injection Control (UIC) permit, gave final approval to start injection of 1,000,000 metric tons of CO_2 into a target zone approximately 7,000 feet deep in the lower Mt. Simon sandstone over a three year period. Injection operations began November 15, 2011, at an average rate of 1,000 metric tons of CO_2 per day. The Illinois State Geological Survey, which leads the project as part of the DOE-supported Midwest Geologic Sequestration Consortium Regional Partnership, estimates the regionally expansive Mt. Simon saline reservoir could store billions of tons of CO_2, offering significant opportunities to store the more than 250 million tons of CO_2 produced each year in the Illinois Basin. The Midwest Geologic Sequestration Consortium Regional Partnership is one of seven NETL-managed regional partnerships studying the viability of carbon storage as a greenhouse gas mitigation strategy for the United States.

Updated National Carbon Sequestration Database and Geographic Information System Launched—The National Carbon Sequestration Database and Geographic Information System (NATCARB)—available at the NETL website—has been updated and redesigned for ease of public access and viewing. The interactive online tool provides a view of carbon capture and storage (CCS) potential in the United States and portions of Canada, and it integrates information on worldwide CCS deployment efforts. Locations of small- and large-scale domestic CCS field projects, including the 10 Recovery Act-funded site-characterization projects, are shown with links to pertinent information. NATCARB was created by NETL with input from the seven Regional Carbon Sequestration Partnerships established by DOE to develop technologies for the safe and permanent storage of CO_2 through geologic and terrestrial approaches best suited to each region.

Midwest Has Potential to Store Hundreds of Years of CO_2 Emissions—Small-scale CO_2 injection field tests conducted by the Midwest Regional Carbon Sequestration Partnership (MRCSP) indicate that the region has the geologic potential to store hundreds of years of CO_2 emissions, primarily in deep saline formations. The MRCSP Phase II field tests included seven small-scale field validation tests consisting of three geologic injection tests—one in each of the major geologic provinces of the region (the Michigan Basin, the Appalachian Basin, and the Cincinnati Arch)—and four terrestrial field tests representative of the region's diversity in

croplands, reclaimed mine lands, reclaimed marshlands, and forested wetlands. The small-scale geologic field tests injected CO_2 into saline formations to validate data gathered in Phase I research. The field tests also found that oil and gas reservoirs have a high potential for enhanced oil and gas recovery.

New Roadmap Updates Status of DOE Carbon Capture, Utilization, and Storage Research, Development, and Demonstration Efforts—DOE published a roadmap that provides an overview of research, development, and demonstration (RD&D) efforts to supply cost-effective, advanced carbon capture, utilization, and storage (CCUS) technologies for coal-based power systems. The "DOE/NETL Carbon Dioxide Capture and Storage RD&D Roadmap" outlines the efforts to develop advanced CCUS technology, as well as several technologies being pursued to mitigate risks inherent to RD&D efforts. DOE anticipates that an array of advanced CCUS technologies will be ready for large-scale demonstration by 2020, providing safe, cost-effective carbon management to meet national goals for reducing greenhouse gas emissions. Research success will enable CCUS

technologies to overcome a wide range of challenges, such as successful integration of CO_2 capture, compression, transport, and storage technologies with power generation systems; effective CO_2 monitoring, verification, accounting, and assessment; permanence of underground CO_2 storage; and public acceptance.

New Study Says State Regulatory Framework Will Likely Result in Robust CO_2 Pipeline System—According to an NETL-funded report, the approach that is most likely to result in a strong CO_2 pipeline system in the United States will be a private-sector model with a state-based regulatory framework. However, the study also notes that a federal role, which would encourage private construction of CO_2 pipelines through incentives, would also be important in the future. The report, titled, "A Policy, Legal, and Regulatory Evaluation of the Feasibility of a National Pipeline Infrastructure for the Transport and Storage of Carbon Dioxide," analyzes a potential pipeline network that would transport the CO_2 emitted from large stationary sources to underground storage locations. A pipeline network is believed to be an important component to commercialize and deploy carbon capture and storage technologies to reduce the buildup of CO_2 in the atmosphere. The report, which was undertaken by the DOE-funded Pipeline Transportation Task Force (PTTF), was developed by SECARB and the Interstate Oil and Gas Compact Commission. The data collected for the report are expected to improve commercialization efforts by analyzing current CO_2 storage situations and identifying what is needed for viable transport to storage areas.

Clean Energy

Carbon Capture, Utilization, and Storage

DOE Best Practices Manual Focuses on Site Selection for CO_2 Storage

DOE released a best practices manual (BPM) focused on the most promising methods for assessing potential CO_2 geologic storage sites. Developed by NETL, the manual, titled, "Site Screening, Site Selection, and Initial Characterization for Storage of CO_2 in Deep Geologic Formations," will continue to be used as a resource by future project developers and CO_2 producers and transporters. In addition, the BPM will be used to inform government agencies of the best practices for exploring potential CO_2 geologic storage sites and to educate the public. This is the fourth BPM released by DOE, and it provides a framework for reporting resources calculated using methods developed by DOE, the Carbon Sequestration Leadership Forum, and others. This BPM focuses on the exploration phase of the site characterization process and communicates analyses and guidelines for narrowing potential sub-regions into qualified sites for CO_2 geologic storage. Development of the geologic storage system proposed in this BPM has been instrumental in establishing consistent, industry-standard terminology and guidelines for communicating storage resources and storage capacity estimates, as well as project risks, to stakeholders.

DOE Manual Studies Terrestrial Carbon Sequestration

According to a best practice manual (BPM) released by DOE, titled, "Best Practices for Terrestrial Sequestration of Carbon Dioxide," considerable opportunity and growing technical sophistication could make terrestrial carbon storage both practical and effective. The BPM details the most suitable operational approaches and techniques for terrestrial CO_2 storage, in which photosynthesis creates organic matter that stores CO_2 in vegetation and soils; this is different from CO_2 mitigation technologies that focus on capturing and permanently storing anthropogenic CO_2 emissions. NETL used data from the seven Regional Carbon Sequestration Partnerships (RCSPs) to prepare the BPM, which also discusses the analytical techniques necessary to monitor, verify, and account for terrestrially stored carbon. In addition, results from the RCSPs' terrestrial field projects are presented. The best practices outlined in this BPM will help those interested in pursuing terrestrial storage projects—and those interested in regulating them—to optimize their efforts.

NETL Technique Characterizes Real-Time Movement of CO_2 Underground

Collaborative research between NETL and the University of Pittsburgh, a participant in the NETL–Regional University Alliance, has produced a technique for characterizing CO_2 movement underground using seismic surveys. The field survey tool is an outcome of correlating the velocities of sonic compression and shear waves in reservoir rocks with their chemical and geomechanical properties following exposure to varying fluid compositional changes. The reservoir rocks examined were obtained from the Scurry Area Canyon Reef Operators Committee field located in north-central Texas—the oldest U.S. CO_2-enhanced oil recovery site, where more than 86.2 million tons of CO_2 were purchased as of 2010. The technique is described in the 2011 Elsevier e-only publication from the 10th International Conference on Greenhouse Gas Control Technologies in 2010, *Energy Procedia* (Vol. 4).

NETL-Regional University Alliance Advances Science of Underground Flow

NETL- Regional University Alliance participants have elucidated the conditions that lead to fractal flow in porous media. Fractal flow can cause a fluid interface to advance more rapidly in certain locations, which is inconsistent with assumptions contained in conventional Euclidean flow models. The researchers have developed and experimentally tested new, theory-based equations to describe the conditions of fluid viscosity, interfacial tension, and other parameters that control the crossover between fractal and Euclidean flow. Results of the study permit the characterization of the capillary number and viscosity-ratio dependencies of various flow descriptors, including the center-of-mass of the injected fluid, its interfacial width, and its saturation and fractional flow profiles in the Euclidean limit. These properties directly affect the efficiency of CO_2 storage and CO_2-enhanced oil recovery, as well as any chemical reactions in the interfacial region. The study is described in the peer-reviewed journal, *Transport in Porous Media* in the January 2011 issue (Vol. 86, No. 1).

NETL Collaborative Research Aims to Optimize CO_2 Mineralization

NETL scientists and Australian collaborators at the University of Sydney and Orica Limited conducted 100 CO_2 mineralization tests on

magnesium-rich rock samples from Australia, Oregon, Pennsylvania, and Washington. The test series explored various experimental parameters including temperature, pressure, time, and slurry density to establish the optimum mineralization conditions for the various ores in an industrial setting. Conversion efficiencies of 50–65 percent were achieved for reaction times of 1–3 hours. This work is intended to provide critical process data on the mineral carbonate product stream, and data that are more definitive on the long-term stability of the solid carbonate product.

NETL-Led Life Cycle Analysis Finds Smaller Carbon Footprint for Synthetic Jet Fuel

A final report issued by the U.S. Air Force (USAF) Research Laboratory demonstrates that synthetic jet fuel can be produced from coal and biomass with a greenhouse gas (GHG) footprint below the petroleum baseline. Peer-reviewed externally, the report is a detailed case study of ten scenarios for producing synthetic paraffinic kerosene for aviation. All scenarios were modeled based on combining the Fischer-Tropsch process with carbon capture for either storage in a saline reservoir, or utilization in enhanced oil recovery. NETL led the analysis as part of the Interagency Workgroup for Aviation Alternative Fuels, a multi-governmental agency, industry, and academia research group convened by the United States Air Force to develop guidance on estimating GHG emissions in aviation applications with currently available data and tools. Section 526 of the Energy Independence and Security Act of 2007 requires that any alternative fuel procured by the federal government be certified as producing fewer emissions than conventional petroleum-based fuels. NETL developed the framework utilized by the Workgroup for incorporating uncertainty analysis when estimating the life-cycle GHG emissions from the production of biomass. The final report and related presentation are available from NETL's public website.

Invention Uses Waste Streams to Sequester CO_2 and SO_2

A number of methods have been tried to mitigate the potential environmental impacts of stored bauxite residue—which is generated when aluminum is extracted from the principal ore called bauxite—in tailing ponds. However, no effective solutions have been found that can deal with the high volume of residues in an economically viable way. The caustic nature of the byproduct presents long-term environmental concerns because leakage of the alkaline liquid from impoundments into groundwater aquifers can result in polluted groundwater. In this patented invention (U.S. patent 7,922,792), NETL researchers answer this need using saline wastewater produced as a byproduct of enhanced oil and gas recovery. With this invention, a mixture of oil or gas wastewater brine and bauxite residue can effectively sequester acid gases. The bauxite residue-brine mixture can neutralize acid gases as well as acid waste material. Additionally, this research found that the mixture is also effective in simultaneously removing both CO_2 and SO_2 from industrial flue gas streams.

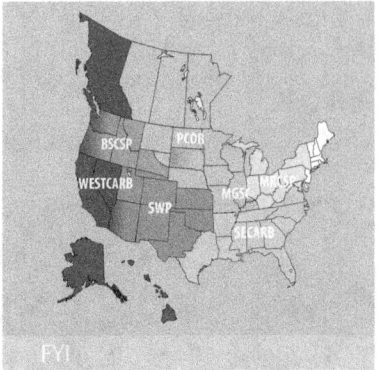

Over 400 state agencies, universities, and private companies in 43 states and 4 Canadian provinces belong to the RCSP (Regional Carbon Sequestration Partnership) program. The RCSP program is developing effective CO_2 transport and storage solutions, as well as advancing public understanding of the process and tailoring these efforts to the needs and resources of its seven regions. NETL leads the Partnerships through management and funding, working groups, technical oversight, and conferences. Currently, CO_2 injection testing is underway for Phases II and III of the program. The RCSP program remains the centerpiece of national efforts to commercialize carbon capture and storage technologies.

Clean Energy

Demand-Side Efficiency

Through NETL's project management, DOE's Office of Energy Efficiency and Renewable Energy (EERE) and its partners work together on technologies that conserve and increase the efficiency of energy consumption in our homes, buildings, vehicles, and in manufacturing.

Long lasting organic LEDs provide highly efficient lighting for kitchen countertops. (Image courtesy of UDC)

New Organic Light-Emitting Diode Under-Cabinet Luminaire Eclipses Planned Performance Expectation—Universal Display

Corporation (UDC), with funding from NETL, has delivered two complete, pre-prototype under-cabinet lighting systems that successfully demonstrated more than 4,200 total lumens with efficiency greater than 60 lumens per watt and color rendering index greater than 85. The luminaires' estimated lifetime is over 20,000 hours. Each delivered system includes 10 OLED panels measuring 15 by 7.5 centimeters, an integrated power supply, special interconnects, dedicated low-voltage wiring, and dimming control. The panels employ proprietary phosphorescent materials and are built with early development versions of UDC's out-coupling, efficiency-improving

technology. Continuing development of advanced out-coupling concepts currently under consideration at UDC are expected to produce even higher system efficiencies with concurrent improvements in the color rendering index.

New Lower-Cost Materials Demonstrated for Organic Light-Emitting Diode Manufacture—

Under an NETL financial assistance research and development award, product developers at Arkema, Inc., and collaborators at Philips Lighting have produced 6-inch square OLED panels made with residential window glass rather than the expensive display-grade borosilicate glass substrates typically used. Moreover, the indium tin oxide that normally coats the substrate to form the OLED anode was replaced with doped zinc oxide that is comparatively plentiful, can be deposited less expensively, and has comparable transparency and conductivity. The use of such low-cost integrated substrates will lead to significantly reduced costs and greater market penetration for OLEDs.

Veeco's economical system for LED production will mean reduced consumer expense. (Image courtesy of Veco Instruments, Inc.)

Improved System Lowers Light-Emitting Diode Manufacturing Cost—The new MaxBright™ multi-reactor system by Veeco Instruments, Inc., partially funded through NETL, will help lower the cost of manufacturing high-brightness LEDs

A new Energy Efficiency Buildings (EEB) Hub has been established. Penn State University is leading a consortium of private businesses, economic development agencies, national laboratories, universities, and community and technical colleges with the aim of cutting energy use in the commercial buildings sector by 20 percent this decade. The NETL-managed project is uniting science and industry to improve energy efficiency and reduce carbon emissions from existing buildings while also stimulating private investment and job creation. The EEB Hub is headquartered at the Philadelphia Navy Yard, one of the nation's largest and most dynamic redevelopment opportunities.

by improving throughput, growth uniformity, yield, and temperature stabilization. Veeco's innovation improves metal organic chemical vapor deposition (MOCVD) technology by offering an advanced wafer carrier design, which holds fourteen 4-inch wafers (compared to the standard 12) to increase capacity. Veeco has also revised the pocket shaping and contoured the wafer holders for better temperature uniformity. The introduction of optimized pocket shaping has resulted in a within-wafer wavelength uniformity improvement of 24 percent. Consequently, the wafer yield for a 5-nanometer wavelength bin has increased from 82 to 92 percent. These changes represent a 24 percent reduction in the cost of ownership for Veeco's MOCVD equipment and will contribute to a significant lowering of manufacturing costs.

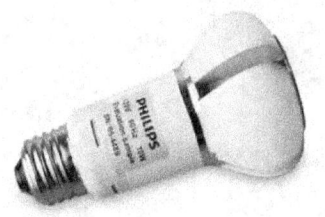

Philips's innovative light bulb shines as brightly as an incandescent bulb but uses a fraction of the energy. (Image courtesy of Philip's)

Philips Product Delivers on DOE Challenge to Replace Incandescent Light Bulb—Philips

Lighting North America was awarded the L Prize for a product designed to replace the 60-watt incandescent light bulb. The LED product consumes under 10 watts to produce an equivalent amount of light but with an energy savings of 83 percent. As the first and only L Prize entrant in the 60-watt replacement category to successfully meet all requirements of the competition, Philips received a $10 million cash prize, with L Prize partner promotions and incentives under development. The winning Philips LED bulb completed 18 months of intensive field, lab, and product testing to meet the rigorous requirements of the L Prize competition that was launched in 2008 and administered by NETL in partnership with Pacific Northwest National Laboratory on behalf of the Office of Energy Efficiency and Renewable Energy. NETL managed the test program, which was performed by independent laboratories and designed to ensure that performance, quality, lifetime, cost, and availability meet expectations for widespread adoption and mass manufacturing. If every 60-watt incandescent bulb in the United States were replaced with the 10-watt L Prize winner, the nation would save about 35 terawatt-hours of electricity or $3.9 billion in 1 year while avoiding 20 million metric tons of carbon emissions.

New Technique Identifies Efficiency-Killing Defects in Light-Emitting Diodes—Researchers at

Sandia National Laboratories participating in an NETL-managed solid-state lighting core technology project have developed a unique approach for identifying defect concentrations and energy levels in LEDs. Defects arise by a variety of mechanisms, such as impurity incorporation during growth or incorrect chemical bonding (when an atom occupies the wrong lattice site in the crystal). These defects can produce non-radiative recombination within LEDs that reduces efficiency by generating heat rather than light. While the importance of reducing defect incorporation in LEDs is widely appreciated, no scientific method had before been available to quantitatively study the energy level or density of defect states in indium gallium nitride and gallium nitride semiconductor materials that make up blue and green light-emitting LEDs used in solid-state lighting. Now, deep-level optical spectroscopy, a technique developed by Sandia researchers, enables the systematic understanding and minimization of defects directly in LED structures, which will result in the development of more efficient LEDs and LED-based light sources. Moreover, the deep-level optical spectroscopy technique can be used to determine the generation of defects over the life of the LED and help researchers to better understand the physical mechanisms that lead to performance degradation over the life of the LED.

New Process Reduces Warm-White Light-Emitting Diode Cost—Current

LED manufacturing technology involves spraying a coating of phosphor onto a glass dome, which results in significant wastage

of the expensive phosphor material. A new process developed at GE Lighting Solutions in cooperation with NETL employs direct injection molding of phosphor domes or inserts in a single manufacturing step. Using the new molding process, the composition of the phosphor blend can be tailored, and the novel process has demonstrated tight control over color temperature and color quality. GE Lighting estimates that the cold deck injection molding process will reduce the amount of phosphor consumed by more than 80 percent, which will have a significant impact on overall manufacturing costs.

DID YOU KNOW

L ight-emitting diodes (LEDs) and organic light-emitting diodes (OLEDs) promise to be 10 times more energy-efficient than conventional incandescent lighting and can last up to 25 times as long. By 2030, these technologies have the potential to reduce national lighting electricity use by nearly one half, which could save up to $30 billion per year.

Clean Energy

Americian Recovery & Reinvestment Act

With funding through the 2009 American Recovery and Reinvestment Act (ARRA), NETL-managed projects have contributed to our domestic economic growth by reaching significant, practical milestones toward cleaner, more efficient energy production and use.

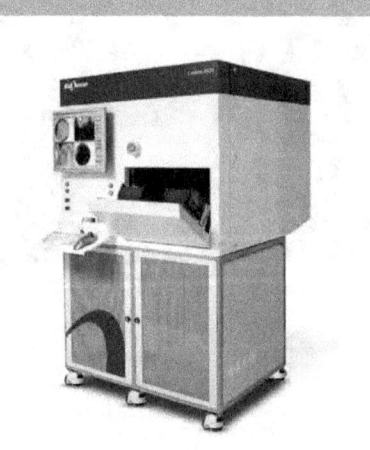

KLA-Tencor's new LED inspection tool detects defects in the early stages of production. (Image courtesy of KLA-Tencor Corp.)

Novel Inspection Tool Reduces Light-Emitting Diode Manufacturing Cost—As part of an NETL-managed solid-state lighting research project, officials from KLA-Tencor Corporation have released an improved inspection tool named Candela® 8620 that promises to significantly improve overall process yields and minimize expensive waste in LED manufacture. The power of the inspection tool lies in optical detection techniques coupled with defect source analysis software to statistically correlate front-end geometric anomalies in the substrate to "killer defects" on the back end of the manufacturing line; these killer defects give rise to an undesirable or unusable end product. The new tool was developed with Recovery Act funding and was beta-tested by a number of key manufacturers, including the Philips Lumileds Lighting Co., where it has been an important part of their yield and cost reduction efforts and an important element in their recent ramp-up from 75 millimeter to 150 millimeter sapphire substrates.

Recovery Act-funded Industrial Carbon Capture and Reuse Project Completes Pilot Plant Design—Collaborators at the Gas Technology Institute, Scripps Institute of Oceanography, and the University of Connecticut concluded work in the use of macroalgae (seaweed) for CO_2 capture and renewable energy by presenting to NETL staff the engineering designs for a pilot-scale facility sized to consume 100 pounds per day of CO_2 from a power plant flue gas stream. The team selected Gracilaria vermiculophylla as the specie to be evaluated in the pilot-scale facility. In laboratory experiments, this red macroalga tolerated a fairly wide temperature swing (78–98 °F), grew on a vertical substrate with only intermittent spraying of water, and harvested easily. Use of G. vermiculophylla in "algal scrubber" technology could provide an effective, efficient, and affordable way to process power plant waste streams into biomass for digestion to methane.

Aerial view of the Black Warrior Basin Gorgas Power Plant site. (Image courtesy of Google Earth, U.S. Geological Survey, and USDA Farm Service Agency)

Recovery Act Project Reports Successful Gorgas Power Plant Well to Target Carbon Storage in Black Warrior Basin—A characterization well sunk 4,915 feet deep on the property of Alabama Power Company's William C. Gorgas 1,400-megawatt coal-fired power plant in Walker County, AL, encountered formations promising for carbon storage with these strata also bounded by competent seals. The extensive borehole data collected included geophysical well logs, core, and seismic test results. Ten miles of 2-dimensional seismic reflection lines were also acquired adjacent to the Gorgas plant to image subsurface conditions. These combined data will be used to determine the extent of stacked storage in the units underlying the Gorgas plant. Hydrocarbon shows were also present in the well at the Fayette sandstone (1,060 feet deep), the Boyles Sandstone (2,027 feet deep), and the Hartselle Sandstone (between 2,622 and 2,716 feet deep). A project team led by University of Alabama researchers is studying the data from the well in cooperation with NETL to more accurately quantify the capacity of Black Warrior Basin saline formations and mature hydrocarbon reservoirs for commercial carbon storage and possible enhanced oil and gas recovery.

The Cedar Lane facility is designed with two indoor and two outdoor algal ponds.

Construction Begins for ARRA Project to Capture and Reuse CO$_2$ from a Coal-Fired Flue Gas Source—Ground breaking began for a pilot plant to convert CO$_2$ into useful products via algae biomass at Cedar Lane Farms, a garden plant company situated on more than 13 acres in Wooster, OH. Developed at Touchstone Research Laboratory (TRL), the algal bioconversion process utilizes a unique phase-change material floating on the surface of an open pond to tackle three significant challenges facing the algae bio-fuels technical community: maintaining temperature, minimizing water evaporation, and excluding invasive species. The TRL process employs novel aqueous-phase technology that extracts oil from the algae without an energy-intensive drying step, recycles nutrients, and produces renewable fuel in the forms of algal oil (biofuel) and residual biomass (methane). Working in cooperation with NETL, the project team produced a topical report on accomplishments to date, including initial analytical results, feasibility analysis, and detailed design and permitting of the pilot plant, construction of which is anticipated to be completed in 2012.

Mikro Systems Surpasses Interim Commercialization Metrics for Innovative Gas Turbine Manufacturing Technology—At the end of fiscal year 2011, product developers at Mikro Systems, Inc., in Charlottesville, VA, reached a significant milestone by requalifying labor and yield values for their proprietary approach to casting ceramic cores for next-generation gas turbine blades. Yield on the validation run was 33 percent, well over the fiscal year 2011 target of 27 percent, and labor time was 205 minutes, under the fiscal year 2011 goal of no more than 220 minutes. Core production cost at commercial volume is now projected to be about half of the baseline cost at project inception. Mikro Systems is applying this patented Tomo-Lithographic Molding (TLM™) manufacturing technology to gas turbine airfoils as part of the Recovery Act-funded, NETL-administered Small Business Innovation Research Phase III Xlerator project. In addition to enabling performance-enhancing designs that were previously impossible to manufacture, the technology will reduce tooling costs and production lead times and also improve manufacturing process efficiency. Siemens Energy entered a collaborative license agreement with Mikro in June 2011 to validate and certify the technology for commercial production of stationary and moving airfoils (vanes and blades, respectively); they will establish a field office near Mikro to more closely support commercialization efforts. Surpassing the interim metrics at the project midpoint reflects Mikro's excellent progress towards commercialization of an innovative technology that will significantly advance the state-of-the-art in industrial gas turbine performance.

First U.S. Large-Scale, Recovery Act-Funded Carbon Capture, Utilization, and Storage Demonstration Project Begins Construction— Working in cooperation with NETL, a project team led by Archer Daniels Midland Company initiated construction activities aimed at demonstrating a process that captures CO$_2$ from an ethanol plant in Decatur, IL, for storage in the underlying Mount Simon Sandstone, a deep saline formation. Because the CO$_2$ is a by-product of fermenting corn into fuel-grade ethanol, this industrial carbon capture and storage project will have a negative carbon footprint. The project team expects to sequester approximately 2,500 metric tons of CO$_2$ per day at depths of 6,500–7,000 feet. The Mt. Simon formation could store billions of tons of CO$_2$, offering significant opportunities to sequester the more than 250 million tons of CO$_2$ produced each year in the Illinois Basin.

Americian Recovery & Reinvestment Act

Aquion's storage system for off-grid solar scenarios offers three times the life of the leading lead-acid battery pack without the maintenance. (Image courtesy of Aquion)

Recovery Act-supported Firm Named Pittsburgh's Startup Company of the Year—Aquion Energy, Inc., developer and manufacturer of ambient-temperature, sodium-ion batteries, was selected by the Pittsburgh Technology Council for the 2011 Startup

Company Tech 50 Award. Each year, the award honors 50 of the Pittsburgh region's fast-growing technology industries and professionals. Somewhat larger and heavier than a car battery, the Aquion energy storage device employs an environmentally-friendly aqueous electrolyte and is aimed at utilities wishing to store electricity for use during peak demand, or to optimize intermittent power producers (e.g., solar, wind). Since receiving $5 million in smart-grid Recovery Act funding from DOE, the firm has attracted an additional $30 million in venture capital and now seeks space for its first high-volume manufacturing facility. The facility is expected to create more than 500 jobs across a wide range of employment categories by 2014, and 1,000 jobs by 2017. NETL manages this and other smart grid projects in support of the research and development program of DOE's Office of Electricity Delivery and Energy Reliability.

Rooftop solar panels increase energy efficiency on homes in Austin, TX. (Image courtesy of Donald Geiling)

Rooftop Solar Energy Production Analyzed—Solar photovoltaic (PV) panels installed on rooftops of some homes in the 711-acre Robert Mueller mixed-use redevelopment project in Austin, Texas, produced from 33 to 90 percent of the kilowatt-hours utilized, depending on the home analyzed. The data were taken for seven homes during August, a very hot month in Texas. The homes are among 172 currently participating in the Recovery Act-funded Energy Internet Demonstration project conducted by Pecan Street, Inc. (PSI) to verify smart grid technology viability, quantify smart grid costs and benefits, and validate new smart grid business models at a scale that can be readily adapted and replicated around the country. PSI also found that energy produced in the afternoon by south-facing panels represented 15.5 percent of energy consumed, but that energy from west-facing panels, albeit less (i.e., 10.9 percent of consumption), came later in the afternoon and closer to peak demand time. Eventually, PSI will install and test

various smart grid technologies in up to 1,000 residential and 25 commercial customers in the Mueller Community. NETL manages this and other smart grid projects on behalf of the Office of Electricity Delivery and Energy Reliability.

Test engine at Kansas State University's National Gas Machinery Laboratory has been used to evaluate new retrofit technologies for emissions control. (Image courtesy of Kansas State University)

Retrofit Technology Improves Performance, Reduces Emissions

—In the natural gas industry, thousands of reciprocating engines are used to produce electricity, compress and re-inject natural gas to increase oil production, or compress natural gas to feed it into gas transmission pipelines. These engines are aging, and new federal Environmental Protection Agency (EPA) emissions regulations may necessitate cost-prohibitive replacements. However, with funding from NETL, researchers at Kansas State University (KSU) and partners are designing and testing retrofit emissions control and monitoring technologies for these engines that will efficiently upgrade them while mitigating greenhouse gases—at a fraction of the

cost of new engine replacements. KSU researchers have developed a four-stroke cycle engine model and exhaust gas oxygen sensor model that will predict emissions from small engines better than current emission compliance measures, which are based on plume models for larger emission sources. The updated KSU models take into account small-engine characteristics and preferred catalytic conditions. Once validated, the models can be used in field engine control boards that can help meet new EPA emission standards by replacing outdated air fuel controllers.

Researchers at Rock Springs, WY, performed tests to evaluate southwestern Wyoming's Rock Springs Uplift as a potential geologic CO_2 storage site. (Image courtesy of the University of Wyoming)

Test Well to Characterize Wyoming Geologic Sequestration Potential

—Working in cooperation with NETL with Recovery Act funding, a project team organized by the University of Wyoming with industry partner Baker Hughes, Inc., completed installation of a stratigraphic test well 12,810 feet into the Rock Springs Uplift and collected attendant geologic, geophysical, and geochemical data and samples. Data collected from the well and other geophysical surveys will allow numerical simulations to yield much more accurate predictions of CO_2 storage capacity and potential plume migration, and the data could be utilized to underpin future large-scale carbon capture, storage, and utilization projects in Wyoming. The Rock Springs Uplift was selected for characterization based on its geologic setting and proximity to some of the state's largest sources of anthropogenic CO_2. Preliminary analyses indicate that the Rock Springs Uplift could store 26 billion tons of CO_2—a capacity sufficient to accept Wyoming's current annual CO_2 emissions of 55 million tons for more than 470 years.

Oil & Natural Gas

Fossil fuels are considered the most dependable, cost-effective energy source in the world. The availability of these fuels to provide clean energy will be essential for domestic and global prosperity well into the 21st century.

Technology development is making oil and natural gas production more economic and environmentally sound.

In the late 1970s, DOE conducted the first scientific assessment of Appalachian shale gas resources and helped accelerate a number of energy technologies that are commonly employed today, such as horizontal drilling and high-volume hydraulic fracturing, detailed natural fracture analysis, and microseismic fracture mapping. The resulting increase in natural gas use has been credited with reviving regional economies and reducing domestic CO_2 emissions to their lowest levels in two decades. From gas hydrate research to enhanced oil recovery, chemical flooding, and research focused on reducing the impact of oil and gas extraction on our environment, NETL is a true pioneer. Looking ahead, methane hydrates could become an important energy resource, and NETL leads the way in studying this resource that is found in ice-crystal structures locked in deep, frigid water.

Simulating the Future of Coal

A Look at NETL's AVESTAR Center

A trainee uses the operator training station to interact with the IGCC dynamic simulator at the AVESTAR™ Center.

Fossil fuel resources, including coal, petroleum, and natural gas, comprise approximately 80 percent of national and international energy production. In the United States, coal-fired power plants alone account for greater than half of the electricity generated. With increasing energy demands worldwide, coal is expected to play a dominant role in meeting future energy needs both here and abroad. But inexpensive energy comes with a tradeoff. While coal may be cost efficient, coal utilization emits CO_2, a greenhouse gas associated with climate change.

Ambitious efforts are underway to advance new technologies for using coal more efficiently, while minimizing the environmental impact of CO_2 emissions. One attractive technology is integrated gasification combined cycle (IGCC)

with CO_2 capture. IGCC power plants use gas and steam turbines to generate electricity. The gas is synthesis gas, a mixture of primarily hydrogen and carbon monoxide produced by a gasifier. Compared to traditional coal combustion power plants, IGCC offers many advantages, including increased power plant efficiency and easier CO_2 removal, resulting in lower-cost electricity. For IGCC power plants to be practically and optimally implemented, though, operators, engineers, and researchers need to be trained on these systems. NETL answered this call with its Advanced Virtual Energy Simulation Training and Research (AVESTAR™) Center, which opened in 2011.

The AVESTAR Center is dedicated to the safe, reliable, and efficient operation of clean energy power plants. To address challenges in achieving operational excellence, AVESTAR brings together high-fidelity dynamic simulators, 3-D virtual immersive training systems, state-of-the-art facilities, and comprehensive training, education, and research programs.

The Center offers three full-scope, high-fidelity, real-time dynamic simulators, including "IGCC with CO_2 Capture," "Gasification with CO_2 Capture," and "Combined Cycle." The IGCC simulator is unique because it merges, for the first time, a gasification with CO_2 capture process simulator with a combined-cycle power simulator. This single, dynamic simulation framework provides maximum flexibility and performance for training and research applications. The simulators train operators, engineers, and researchers on normal and faulted IGCC plant operations, as well as plant startup, shutdown, and power demand load changes. The system also allows users to test different types of fuel sources and co-firing fuel mixtures.

Throughout 2011, the AVESTAR team and its partners continued to develop an immersive training system (ITS) for IGCC power plants with CO_2 capture. Using virtual reality technology, the ITS provides a 3-D plant walk-through environment. When combined with the IGCC dynamic simulator, the ITS will enable control room and field operators to coordinate their activities and perform collaboratively as a team. The IGCC ITS is scheduled for deployment at the AVESTAR Center in summer 2012.

Since AVESTAR's inception, NETL has continued to build its portfolio of dynamic simulators and operator training systems. Current projects are designed for future integration with post-combustion carbon capture and compression systems, including—

- A natural gas combined cycle dynamic simulator, which will be an offshoot of the combined cycle portion of AVESTAR's IGCC dynamic simulator. The AVESTAR research and development team has already made significant progress by completing plant design and generating an initial dynamic model.

- A supercritical once-through pulverized coal dynamic simulator, which will serve as the baseline power plant for DOE's Carbon Capture Simulation Initiative, aimed at using modeling and simulation tools to accelerate the commercialization of carbon capture technologies.

With its portfolio of power plant dynamic simulators and CO_2 capture interfaces, the AVESTAR Center will be well positioned to serve as a Virtual Carbon Capture Center—a concept that will enable AVESTAR's industry partners to test and optimize the operation and control of post-combustion CO_2-capture dynamic models when integrated with baseline power plant dynamic simulators.

By providing comprehensive training and innovative collaborative energy research opportunities, NETL's AVESTAR Center has united state-of-the-art dynamic simulators, virtual reality technology, and dedicated energy experts. This Center is ahead of the curve for clean-energy technologies, and its initiative to accelerate the widespread commercial deployment of such technologies has increased efficiency and reliability, enhanced safety, and reduced environmental impacts.

Background image: A virtual plant that can be used to practice training routines for malfunctions and emergencies that could not be attempted in the real plant, like leaks or fires.

In 2010, NETL demonstrated carbon dioxide flooding EOR in the Hall-Gurney field in Russell Kansas, using carbon dioxide recovered from a nearby ethanol plant. (Image courtesy of Martin Dubois, Kansas Geological Survey)

Project Demonstrates Improved Oil Recovery with CO_2 Storage—

Investigators at the University of Kansas Tertiary Oil Recovery project have established the feasibility of near-miscible CO_2 flooding for improved oil recovery from reservoirs in the Arbuckle Formation of Kansas. The study showed that injection of CO_2 below the minimum miscibility pressure (MMP) can also improve incremental oil recovery. Core examinations verified greater than 50 percent recovery of residual oil remaining after water-flooding when CO_2 was used to displace Arbuckle oil at pressures below the MMP. Near-miscible CO_2 flooding may avoid abandonment of up to 6,407 mature, highly compartmentalized Kansas oilfields, approximately one-third of which have an average of 5 producing wells. The approach could recover 250 to 500 million barrels of incremental oil, significantly extending the life of these oilfields. The project is administered by the Research Partnership to Secure Energy for America, under contract to NETL.

Lateral Well Improves Oil Recovery in Kansas Pilot Test—

Detailed reservoir characterization at a pilot site in the Kansas Unger oilfield using the latest well log data prompted the drilling of a lateral well at a depth of 1,100–1,500 feet. Oil production from this lateral well was 25 barrels of oil per day compared to 3 from the vertical well. Based on this success, American Energies Corporation, a small producer, plans to drill many more wells to boost production from the area. In 2009, 17 wells in the area produced 16,191 barrels of oil at the rate of 2.9 barrels per day. A new enthusiasm to improve oil recovery by drilling lateral and horizontal wells has spurred aggressive lease purchasing by various medium and small operators in Kansas. The project was funded through the EPAct 2005 Section 999 Program.

Microbial Flooding Enhances Oil Production in Mississippi—

When the microbial permeability profile modification (MPPM) procedure— developed at Mississippi State University in cooperation with NETL—was coupled with CO_2 flooding at the Upper Cretaceous Little Creek Oil Field situated in Lincoln and Pike counties, a significant increase in production occurred for two of the five test wells. The study determined that by utilizing environmentally friendly nutrient solutions to stimulate growth of indigenous microflora in the most permeable zones of the reservoir, MPPM technology diverted CO_2 to less permeable, previously unswept zones, thus increasing oil production without interfering with the CO_2 flooding operation.

NETL Nanoadditive Could Customize Water-Based Drilling Mud—

NETL researchers have synthesized anionic clay nanoparticles by pulsed laser ablation in water. Adding a minute amount (around 0.01 weight percent) of the material to a host clay-in-water (1 weight percent laponite) colloidal suspension results in dramatic changes in rheological properties of the mixture. The mixture quickly becomes less viscous (with some memory effects) when a constant shear rate or shear stress is imposed, and gradually recovers viscosity once it is removed. Rheology, the study of the flow of matter, is extremely important to drilling muds, which must cool and clean the drill bit and lift solids to the surface. Results of the investigation appear in the May 2011 issue of the peer-reviewed Elsevier publication *Polymer* (Vol. 52, No. 10).

Technology Attracts Industry Interest—

A low-cost, wet gas compressor developed at laboratory scale by OsComp Systems (Cambridge, MA) with NETL support through the Stripper Well Consortium has attracted more than $12 million for field tests and further development—four times the original goal. The innovative technology lowers capital costs, is capable of a 42:1 compression ratio, has both wet gas and multiphase compression capability, uses less fuel gas, and dramatically reduces the footprint of compression operations. In addition to its use for stripper wells, the compression technology has applications to small-scale liquid natural gas, enhanced oil recovery, mobile compression, sour aggressive gas, and for the compressed natural gas refueling industry. This has led to discussions with Saudi Aramco, which is interested in purchasing multiple larger-scale compressor units for use in its fields.

NETL Releases Report on Life Cycle Greenhouse Gas Inventory Related to Natural Gas—

A new NETL report and presentation titled "Life Cycle Greenhouse Gas Inventory of Natural Gas Extraction, Delivery, and Electricity Production" concludes that natural gas-fired baseload power production has life cycle greenhouse gas (GHG) emissions 42

to 53 percent lower than those for coal-fired baseload electricity, after accounting for a wide range of variability and compared across different assumptions of climate impact timing. The research details the life cycle greenhouse gas emissions from six domestic sources of natural gas (onshore, offshore, associated with oil production, tight sands, shale gas, and coal-bed methane) and a national average mix for extraction and delivery to a large end-user. The report also compares the use of natural gas for power production to coal-fired power production based on the delivery of 1 megawatt hour of electricity to the end user. Lower GHG emissions for natural gas result primarily from differences in the current fleet's average efficiency (53 percent for natural gas versus 35 percent for coal) and higher carbon content per unit of energy for coal compared to natural gas.

NETL Researchers Make First-Ever Observation of Methane Hydrate Phenomenon—While gas hydrates are predicted to reform around a wellbore during depressurization-based gas production from gas hydrate-bearing reservoirs, the phenomenon had never been observed. Now, utilizing a medical x-ray CT scanner, NETL researchers captured images of gas hydrates reformation while producing gas from a hydrate-bearing sand sample in a pressure vessel by depressurizing the container with and without heat addition. CT images of the hydrate-bearing sample were processed to provide 3-D data of heterogeneous porosity and phase saturations suitable for numerical simulations. In the experiments, gas hydrate reformation was observed only in the case of no-heat supply from surroundings—a finding consistent with numerical simulation. The study is detailed

in the March 2011 issue of the American Chemical Society publication, *Energy & Fuels* (Vol. 25, No. 3).

The drilling rig on the site of the DOE-ConocoPhillips "Ignik Sikumi" gas hydrate field trial well in the Prudhoe Bay Unit (Alaska North Slope). In 2011, the well confirmed the presence of multiple underground reservoirs bearing gas hydrate, collected detailed geologic information, and installed an array of down-hole well monitoring devices in advance of production testing operations scheduled to occur in early 2012.

Massachusetts Institute of Technology Report Recognizes National Methane Hydrate R&D Program Successes—A major study released by the Massachusetts Institute of Technology (MIT) on the future of natural gas is strongly positive regarding the potential benefits of gas hydrate research and development and is highly complimentary of the accomplishments of the National Methane Hydrate Research & Development Program. The DOE-led program is a collaboration of seven federal agencies with a broad mandate to improve the understanding of gas hydrate's role in the environment and potential as a source of natural gas. Work conducted in cooperation with NETL under the program is cited throughout the gas hydrate discussion, which recognizes numerous NETL successes attained through

partnerships with other government agencies, academia, and industry in field, modeling, and laboratory programs. The report notes that because production data are limited, the long-term possibility of methane production from hydrates remains unproven, and recommends that U.S. government sponsorship be sustained, particularly for research aimed at demonstrating production feasibility and economics.

Hydrates Confirmed at Gulf of Mexico Site—One of the five jumbo piston cores obtained at the site of the Mississippi Canyon Block 118 seafloor observatory was found to contain hydrates in the form of massive chunks, blades, nodules, and disseminated grains. The core is the first confirmation of subsurface gas hydrate at the site. The finding supports prior geophysical interpretations that suggest that gas delivery to the seafloor, and the distribution of subsurface gas hydrate, are primarily controlled by faults. The cores were logged at the Stennis Space Center and partially analyzed by scientists from the University of Southern Mississippi and the University of South Carolina. Results of these analyses will advance the geologic, geophysical, and geochemical characterization of the shallow seafloor, and determine sites for an extensive coring expedition scheduled for September 2012. The cores were obtained by a team of researchers participating in the Gulf of Mexico Hydrate Research Consortium led by the University of Mississippi Center for Marine Resources and Environmental Technology with co-sponsorship from NETL and agencies within the U.S. Departments of Commerce and Interior.

In cooperation with DOE, USGS researchers deploy a mini-sparker source to image seafloor sediments in Alaska's Beaufort Sea .

Gas Hydrate Expedition Initiates in U.S. Beaufort Sea

As part of the National Methane Hydrate R&D Program implemented by NETL, scientists with the U.S. Geological Survey (USGS) at Woods Hole, MA, embarked August 2, 2011, on a 10-day expedition to acquire high-resolution geophysical data on the shallow part (less than 20 meters of water depth) of the Central U.S. Beaufort Shelf. The science team applied sonar and seismic techniques to survey the seafloor and water column, collected short vibracore samples (technique for core-sampling underwater sediments) from gas-charged and gas-free sediments, and sampled surface water for post-expedition analysis. Compiled with legacy USGS seismic data and proprietary industry data, results from the expedition complete a first-ever regional map of subsea permafrost distribution on the entire U.S. Beaufort Shelf. The data also test the hypothesis that—unlike other margins such as the East Siberian Arctic Shelf where methane super-saturation observed in the water is considered at least partly linked to active degassing of methane hydrate deposits—most of the gas hydrate associated with subsea permafrost on the Beaufort margin has dissociated since the onset of thawing caused by Late Pleistocene warming and sea level rise.

Zero Discharge Water Management Demonstrated for Horizontal Shale Gas Well Development

A 150-gallon-per-minute mobile treatment unit (MTU) designed by researchers at West Virginia University and FilterSure, Inc., in cooperation with NETL, successfully treated 280,000 gallons of Marcellus shale fracture flowback water while supporting a 9-stage fracture treatment conducted by Chesapeake Energy at a site outside Wheeling, WV. The five-stage mobile treatment unit operated without any operational issues or unit malfunctions, automatically backwashing the filter media every 12 hours. Of the 280,000 gallons of flowback water treated, 98.6 percent (or 276,080 gallons) was returned to Chesapeake Energy for reuse during subsequent fracture treatments. Although the long-term effects on production resulting from the use of recycled water for hydraulic fracturing are not yet known, operators are working to reuse nearly all flowback water for succeeding operations. Successful deployment of the MTU will help reduce the need for fresh water in the fracturing process, thus lowering the environmental impact of gas shale operations.

Project Aids Compliance with New Mexico Waste Pit Regulations

Researchers at the Petroleum Recovery Research Center have developed a web-based geographic information system (GIS) to help oil and gas operators comply with revised rules regulating New Mexico oil and natural gas waste pits. The GIS layers of the New Mexico Pit Rule Mapping Portal are presented for a number of relevant factors including subsurface water depth, surface geology, and soil maps. Base layers for topographic data and aerial photographs allow nearby structures to be located. The project was completed with wide support from both industry and government as part of DOE's Ultra-Deepwater and Unconventional Natural Gas and Other Petroleum Resources Research and Development program. Administered by Research Partnership to Secure Energy for America under contract to NETL and funded from lease bonuses and royalties paid by industry to produce oil and gas on federal lands, the program is designed to increase supply and reduce costs to consumers while enhancing global U.S. leadership in energy technology through development of domestic intellectual capital.

NETL-Developed Process Allows Industrial Use of Produced Water

Researchers at New Mexico Institute of Mining and Technology, working in cooperation with NETL, successfully developed and field-tested a process that, by combining an available pretreatment technology with forward osmosis technology, converts produced water from oil and natural gas wells into a resource that can be used by other industries. The Lea County government, project partner, is actively raising $1.5 million to scale up the 1,000-gallon-per-day design to 100,000 gallons per day. A method for making produced water suitable for mining and to prepare potash for market is timely in view of a recently discovered deposit in arid southeastern New Mexico that could demand an estimated 8 million gallons per day of water for processing.

Snow Fence Supplements Water Supply from Arctic Lake—Researchers at the University of Alaska Fairbanks, working in cooperation with NETL, have shown that installation of a snow fence added approximately 2 million gallons of water to an experimental lake on the Arctic slope during the open-water season. Based on total expenditures for snow fence materials, installation, and labor, the cost of the additional water is 1.2 cents per gallon compared to 13 cents per gallon for water delivered from a nearby source. Because each mile of ice road typically requires approximately a million gallons of water, using snow fences could increase lake recharge and lower the cost of ice road construction. Ice roads are preferred for oil and gas drilling operations because, unlike gravel roads, ice roads leave behind little or no trace, and require no mitigation or reclamation activities when no longer used.

Small Business Innovation Research Yields Breakthrough in Treating Produced Water—With funding from an NETL-administered Small Business Innovation Research (SBIR) grant, product developers at Ohio-based ABSMaterials, Inc., have developed and tested a technology to remove hydrocarbons from flow-back and produced waters resulting from hydraulic fracturing. Employing Osorb®, a sol-gel derived material that rapidly swells up to eight times its dried volume upon exposure to non-polar liquids, the technology effectively absorbs dissolved hydrocarbons and organic acids that make up a majority of oil and gas waste streams. Swelling reverses when the material is heated to evaporate absorbed species. Two pilot-scale Osorb-based

water treatment systems—a 4-gallon-per-minute, non-regenerating, skid-mounted system, and a 60-gallon-per-minute, trailer-mounted system—have been tested with numerous water samples of flow-back water from the Marcellus, Woodford, and Haynesville shale formations and produced water from the Clinton and Bakken formations. A major oil services company field-tested the technology with produced water from the Clinton formation in Ohio in July 2010 and March 2011. The novel technology is gaining the attention of global energy companies and has potential for significantly reducing the environmental impact of producing natural gas from the Marcellus shale and other geologic formations.

Marcellus Shale Hydraulic Fracture Water Treatment Initiated—A project team led by Altela, Inc., working with NETL has installed a hydraulic fracturing flow-back water treatment system at a well site in Indiana County, PA. The project team, consisting of Altela, Inc., CWM Environmental, Inc., and BLX, Inc., have successfully deployed Altela's patented mobile desalination system to treat hydraulic fracturing flow-back and produced water directly at the wellhead in the Marcellus shale. The Altela water treatment system removes all contaminates from the flow-back water generated from hydraulic fracturing and natural gas production. To date, approximately 105,000 gallons of Marcellus shale hydraulic fracturing flow-back water have been treated and purified at the well site, resulting in the production of approximately 79,800 gallons of clean distilled water.

What is shale gas?

During the Devonian period, fine silt and clay particles were deposited in ancient swamps that, with time and pressure, created shale. Plants and animals were buried, too, creating methane. Some methane escaped into rock layers adjacent to the shale, but most of it, which we now call shale gas, remained locked in the shale layers.

Environmental Impact Reduction

AltelaRain® 600 unit module and water treatment towers.

U.S. gas production from shale, including the Marcellus shale, has increased dramatically over the past decade. Tapping this resource with hydraulic fracturing—using pressurized liquids to fracture subsurface rock—and other techniques pioneered by NETL and its research partners has become increasingly important in domestic oil and natural gas production. The water resources needed to hydraulically fracture the Marcellus shale and the potential effects of hydraulic fracturing on surface and subsurface water sources have become key concerns for state legislatures, land owners, and the public. Working with industry and academia, NETL is addressing these concerns by funding multiple projects to develop environmental tools and technologies that will improve management of water resources, water usage, and water treatment during oil and gas exploration and production. These projects could solve some of the most vexing shale gas water problems. In 2011, two of these projects met with notable success: AltelaRain® and Osorb®.

As part of NETL-sponsored demonstration, Altela Inc.'s AltelaRain 4000 water desalination system was tested at BLX, Inc.'s Sleppy well site in Indiana County, PA, for its ability to turn wastewater from shale gas production into distilled water. All of the clean water produced at the demonstration site was suitable for beneficial re-use by well operators for additional simulations and was suitable to be discharged to surface waterways, which reduces the economic and environmental impacts of clean water usage. As a result of the project, Altela designed larger towers for the system, and four AltelaRain 600 modules were sold and installed in Williamsport, PA, to treat approximately 100,000 gallons per day of produced and flowback water from hydraulic fracturing. This represents the first of many planned facilities to be developed in the Marcellus shale basin and similar shale gas basins throughout the United States.

The project is proving particularly useful for Marcellus shale drilling, as evidenced by Altela's partnership with Casella Waste Systems, Inc., which created Casella-Altela Regional Environmental Services, LLC (CARES). The partnership will use AltelaRain to solve environmental issues surrounding the treatment of waste water from Marcellus shale drilling. The Casella-owned landfill in McKean County, PA, will not only be the treatment location, but it will also provide power for the system from clean energy generated by methane gas captured from the landfill. It's a double environmental win by cleaning the water and reducing greenhouse gas emissions. Water will be transported to and from the site via an adjacent railroad, minimizing truck traffic in the area.

ABSMaterial's Osorb technology is another NETL project that will improve management of water resources, usage, and treatment during oil and gas exploration and production. This novel technology uses swelling glass to remove impurities from water and has been shown to clean produced water and flowback waters from hydraulically fractured oil and gas wells. Produced waters, containing a wide variety of hydrocarbons and other chemicals, are by far the largest volume byproduct associated with oil and gas exploration and production.

Two pilot-scale Osorb-based water treatment systems have been built to date: a 4-gallon-per-minute non-regenerating skid-mounted system and a 60-gallon-per-minute trailer-mounted system that includes a mechanism for Osorb regeneration. In independent testing, the skid-mounted system removed more than 99 percent of oil and grease, more than 90 percent of dissolved BTEX (benzene, toluene, ethylbenzene, and xylenes), and significant amounts of production chemicals from the water. Testing of the trailer-mounted system on produced water streams slashed total petroleum hydrocarbon levels from 227 milligrams per liter to 0.1 milligrams per liter. The results of this project have garnered commercial interest in future collaborative efforts from several global energy companies.

A number of other projects are in the works for removing impurities from produced waters through NETL and our partnerships. These projects hold promise for the future of shale gas production, allowing us to tap this important, abundant national resource while protecting the plants, wildlife, and human health.

Background image: Viewed through a fluorescent microscope, the presence and type of bacterial cells in Marcellus shale flowback water was confirmed. Cells are stained with a fluorescently-labeled DNA probe.

A Legacy of Benefit

The Return on Federal Research at NETL

In the world of technology innovation, both the private and public sectors play critical roles in bringing scientific and engineering solutions to American markets.

Breakthroughs that have revolutionized the energy industry have often been made possible by our collective investment in the ideas of the researchers who are tackling problems presented by the production and use of fossil fuels, the generation of nuclear power, the promise of renewable energy, and our need to continuously improve energy efficiencies. Federal research often pursues high-risk technologies that have the potential for great public benefit and require long-term investments the private sector can find difficult to make. It also helps industry bring novel innovations to commercial readiness by sharing in demonstration-scale research.

In 2011, the Office of Fossil Energy took a look at the many public benefits that have been realized by technologies and processes developed and supported by its office and NETL.

Did you know the United States has reduced its NO$_x$ emissions 88 percent and SO$_2$ emissions 82 percent since 1970, essentially eliminating acid rain?

Major contributors to these reductions have been the scrubbers, low-NO$_x$ burners, and selective catalytic reduction systems demonstrated through the clean coal programs managed by NETL. Continued research is expected to directly realize another 37 million ton reduction in SO$_2$ emissions and 16 million ton reduction in NO$_x$ emissions between 2000 and 2020, while also reducing other pollutants.

Did you know that low-cost mercury controls developed at NETL are being deployed on U.S. coal-fired power plants?

NETL's Mercury Control program research reduced the cost of controlling the mercury emitted by coal power systems by 50–70 percent. As of 2010, vendors have sold nearly 150 full-scale activated carbon injection (ACI) systems, a signature technology of the program, to U.S. coal-fired power generators. This represents more than 56 gigawatts of our coal-fired electric

generating capacity. The U.S. Environmental Protection Agency projects that 60 percent of our coal capacity (146 gigawatts) will use ACI by 2015.

Did you know that NETL has been a pioneer in enhanced oil recovery for nearly a century?

Since the 1920s, NETL and its predecessors have pioneered enhanced oil recovery (EOR) techniques like water flooding, chemical applications, and CO$_2$ injection. Coaxing hard-to-recover oil out of the ground adds to our domestic petroleum resources—today enhanced recovery amounts to nearly 13 percent of total U.S. oil production. CO$_2$-EOR, especially if next-generation technologies are developed and deployed, promises a profitable avenue for sequestering manmade CO$_2$ emissions deep underground.

Did you know that U.S. shale gas production has increased 12-fold over the last decade?

Advanced technologies have helped increase oil and natural gas production from abundant shale resources. NETL helped

lead the characterization of shale plays and development of such techniques as horizontal drilling, microseismic monitoring, and hydraulic fracturing, making it possible for us to tap widely present shale layers and release the natural gas trapped inside. With these technologies comes the potential for the United States to move closer to energy independence and even export natural gas.

Did you know the Clean Coal Technology program, implemented by NETL, is expected to realize a $13 benefit for every taxpayer dollar invested?

In fact, between 2000 and 2020, the clean coal program is expected to realize $111 billion in total monetary benefits. Our air will be cleaner, we'll see further reductions in pollution, energy efficiencies and U.S. exports will increase, and national security will be enhanced. The program is also expected to generate 1.2 million jobs, predominantly well-paid manufacturing-related employment.

Like our nation's domestic energy supplies, American taxpayer dollars are a precious resource. They must be carefully managed and they must yield real benefits. From sustainable coal technologies to unconventional petroleum recovery and beyond, the extraordinary economic returns the Office of Fossil Energy and NETL have realized on the investments made through our organizations is a testament to the productivity of our research.

The risks are sometimes high, but so is the return. Our researchers and our partners will continue to work ahead of the curve, and we will continue to achieve the technological breakthroughs needed to keep our nation's energy industry on the cutting edge.

To see the Office of Fossil Energy's complete Legacy of Benefits series, visit them on the Web at **www.fossil.energy.gov**

$1.9B Federal

$3.4B Private Sector

Total cost of the original Clean Coal Technology Demonstration program was $5.3 billion. The cost-share requirement was a 50/50 split, but private sector confidence in the program was high enough to spur a 2/3 investment.

Awards

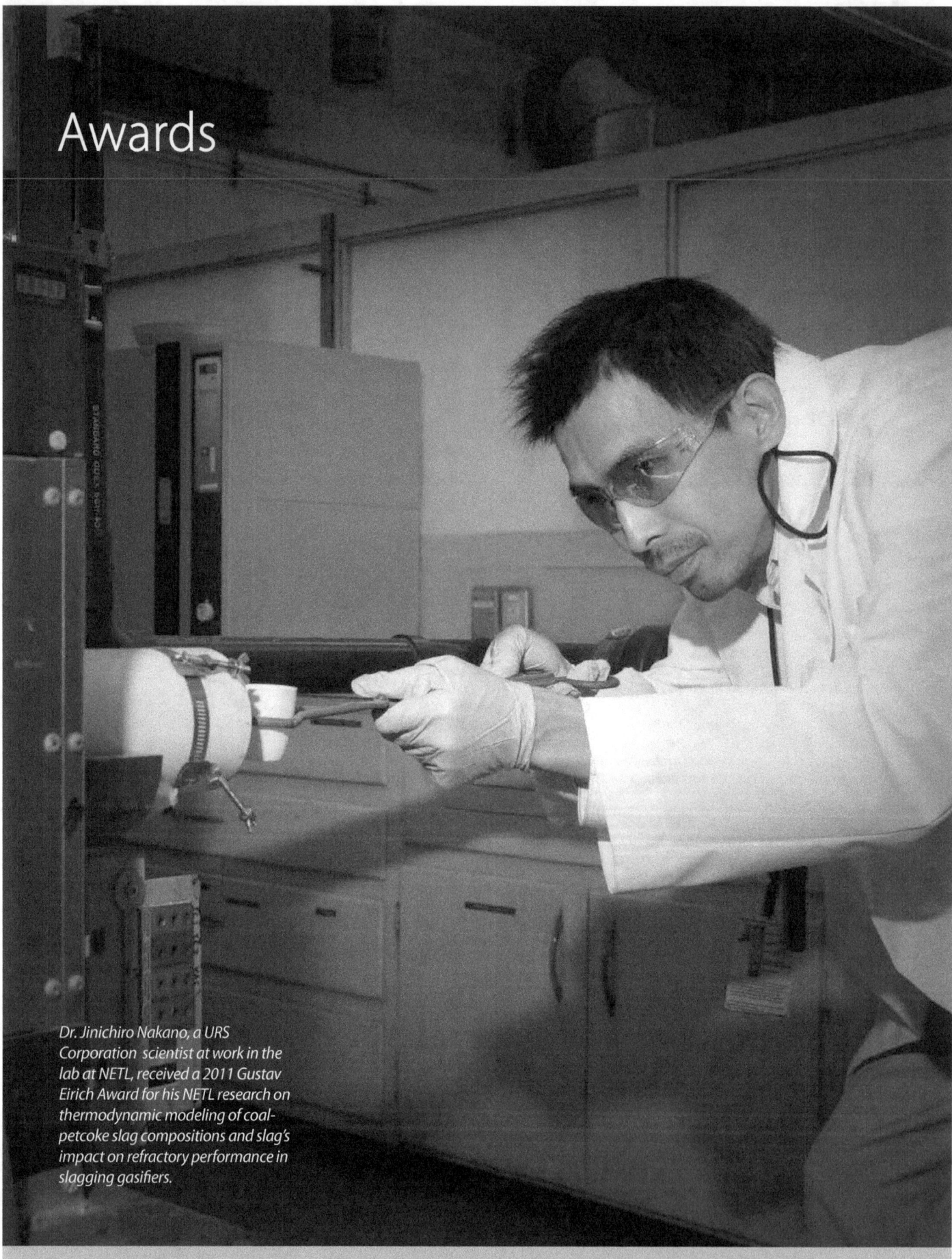

Dr. Jinichiro Nakano, a URS Corporation scientist at work in the lab at NETL, received a 2011 Gustav Eirich Award for his NETL research on thermodynamic modeling of coal-petcoke slag compositions and slag's impact on refractory performance in slagging gasifiers.

Achievement awards recognize NETL's researchers and their technological developments.

NETL has enjoyed a long history of garnering recognition for research excellence, technology transfer, and effective partnerships. Over the past 5 years, NETL's scientists have earned 14 R&D 100 Awards, as well as 14 regional and national awards for excellence in technology transfer from the Federal Laboratory Consortium. These awards, along with the many individual awards given to our scientists and research partners, affirm NETL's contribution toward bettering our nation's future. We're pleased that several of our researchers earned recognition in 2011, from Engineer-of-the-Year to Activated Carbon Hall of Fame, and we feel confident that our lab will continue to be recognized as a world-class research center of excellence, making great strides toward a better future for all Americans.

Awards

NETL Scientist Wins 2011 Gustav Eirich Award
—Dr. Jinichiro Nakano received a 2011 Gustav Eirich Award based on research conducted by NETL on thermodynamic modeling of coal-petcoke slag composition and its impact on refractory performance in slagging gasifiers. Established to recognize outstanding young researchers in the field of refractory materials development, the award was presented to Dr. Nakano at the 54th International Colloquium on Refractories. Dr. Nakano is the first U.S.-based researcher to win this award.

E-Commerce System Wins Award
—The Council of State Governments recognized the Colorado Gas Conservation Commission (COGCC) for developing an eForm permit system that

increases the efficiency of data transfers to help protect groundwater resources. The Colorado eForms system won a GOGCC Innovation Award in the Natural Resources program category. Co-developed with the Ground Water Protection Council in cooperation with NETL, the system reduces permit processing time between state agencies and industry operators by allowing oil and gas operators to complete regulatory forms online. The system streamlines the permitting process by allowing regulatory review to proceed while the public views and comments on applications. The eForm system is also utilized in Nebraska and Alabama, and it is being considered for implementation in Pennsylvania, Kentucky, and Montana.

NETL Researcher Inducted to Activated Carbon Hall of Fame
—Evan Granite received the 2011 Activated Carbon Hall of Fame Award at the 28th International Activated Carbon Conference held in Pittsburgh, PA, October 6–7, 2011. The Conference Director cited Dr. Granite for his research on capturing the trace elements mercury, arsenic, selenium, and phosphorus from coal-derived gas streams. The award is a positive reflection on work

completed by NETL on the cleaning of coal-derived gas streams. The estimated market for recovering mercury from electric power plants and municipal incinerators is $1 billion, according to Professional Analytical and Consulting Services, Inc., the Conference sponsor.

Carbon Storage Leadership Forum Honors DOE-Supported Project
—The Carbon Storage Leadership Forum (CSLF) recognized the International Energy Agency Greenhouse Gas (IEAGHG) Weyburn-Midale CO_2 Monitoring and Storage project with a Global Achievement Award for its role in pioneering carbon capture, utilization, and storage (CCUS) research at commercial operations in Canada. In presenting the award at its ministerial meeting in Beijing, China, on September 21, 2011, the CSLF heralded the project as one of three winners that exemplify global cooperation in implementing CCUS technologies while accommodating scientific research. Over the past 11 years, the project has hosted scientists from more than 40 independent organizations around the globe to study the CO_2 enhanced oil recovery (EOR) and storage operations in the oil fields of

southern Saskatchewan. To date, more than 20 million metric tons of CO_2 have been stored about a mile deep in the carbonate reservoir while dramatically increasing oil production. The amount represents the single greatest quantity of CO_2 stored anywhere in the world, and is equal to the amount exhausted from four million cars in a year. The Petroleum Technology Research Centre located in Regina, Saskatchewan, manages the project with sponsorship from DOE, Canada's federal and provincial governments, the IEAGHG Research and Development Programme, Japan's Research Institute of Innovative Technology for the Earth, and ten corporate sponsors in Canada, Europe, the Middle East, and the United States. NETL represents DOE's interest in this CCUS field study to determine the permanency and best practices of geologic carbon storage in conjunction with the world's largest CO_2 storage and EOR operation.

NETL Researcher Honored at Society of Petroleum Engineers Conference—NETL researcher Dustin McIntyre received a best paper award from the Young Professionals group of the Society of Petroleum Engineers

(SPE) at the 2011 SPE Annual Technical Conference and Exposition in Denver, CO. Dr. McIntyre's paper, "Analysis of Calibration Materials to Improve Dual-Energy CT Scanning for Petrophysical Applications," discusses the use of inductively coupled plasma atomic emission spectroscopy and laser-induced breakdown spectroscopy to determine impurity levels in calibrations standards used when employing dual energy computer tomography to scan coal samples.

NETL Technology Transfer Achievement Recognized—The Federal Laboratory Consortium presented a 2011 Excellence in Technology Transfer Award to NETL for successfully making a new commercially relevant dry sorbent CO_2 capture technology available to the marketplace. NETL entered a cooperative research and development agreement with ADA Environmental Solutions to develop regenerable basic immobilized amine

sorbents that selectively extract CO_2 from flue gas and, upon being heated separately, release the CO_2 for storage. The award was presented at a ceremony during the Consortium's national meeting in Nashville, TN.

NETL Scientist Receives Engineer-of-the-Year Award—The National Association of Corrosion Engineers (NACE) selected Gordon Holcomb as the 2011 Western Area Engineer of the Year for outstanding contributions to both fundamental and applied corrosion science. The honor recognizes the significance of Dr. Holcomb's work in corrosion and corrosion control of infrastructure, atmospheric corrosion of metals and alloys, corrosion detection in pipelines, development of high-temperature corrosion sensors for fireside applications, fundamental research on high-temperature oxidation and hot corrosion, and development of materials for advanced energy systems operating with ultra-supercritical steam and oxy-fuel combustion environments. The award was announced at the 2011 NACE Western Area Conference held November 9–11 in San Diego, CA.

Technology Transfer & Publications

A backscatter electron image from a scanning electron microscope shows the commercial purity of titanium made by International Titanium Powder using the Armstrong Process.

10μm

The crucial step to ensure that our inventions and technologies serve the greatest number of people.

Serving the public and providing a return on investment for federal funding is central to NETL's mission. Our highly skilled researchers tackle energy's toughest challenges through basic research and effective partnerships with industry, academia, not-for-profit, and other government agencies. We recognize that transferring our findings and energy innovations to the public domain is vital for these breakthroughs and innovations to return the most benefit to our nation in the form of improved technologies and real economic impact. NETL meets these objectives through partnerships, patents, licensing, publications, and policy discussions—efforts that garnered four technology transfer awards in 2011 alone. Our scientists and engineers are also active in publishing their findings in print, over the Internet, and through NETL's website. By sharing the knowledge uncovered from our research, NETL accelerates technology developments that support a sustainable energy future.

Power Production With Less Pollution

A close-up of a monolith structure that will be coated with NETL's pyrochlore catalyst to reform diesel fuel into syngas.

Remember the days when starting a car caused a great cloud of exhaust to burst from the tailpipe? Remember the smell? We have come a long way in the past few decades toward reducing the environmental effects of our transportation, but fuel efficiency and pollutants emitted during fuel combustion are still major concerns for the energy industry. NETL's pyrochlore catalysts address both issues.

The Lab's research in this area focuses on converting heavy hydrocarbons, such as diesel and coal-based fuels, into hydrogen-rich synthesis gas. This "syngas" is used in fuel cells and other power applications and can lead to energy technologies that are more efficient

and less polluting. Current reforming catalysts increase the chemical reaction's speed and efficiency, but the high sulfur and aromatic content of these fuels can deactivate the catalyst, requiring that it be replaced.

This is where the pyrochlore catalysts come in. Not only do these catalysts produce more reliable syngas than any other catalysts currently on the market, but they produce it at a potentially lower cost. NETL researchers have incorporated small quantities of rhodium, a precious metal you might find coating a "white gold" ring, into the pyrochlore crystal structure and applied the catalyst to a support. This system reforms an array of fuels into syngas to power solid oxide fuel cells that can produce electricity for transportation, homes, businesses, schools, hospitals, and military applications.

Fuel cells produce electricity with twice the efficiency of conventional systems while conserving resources and reducing pollution. They have been used reliably for decades in space missions where compact, highly efficient power generation is critical. They have many other applications as well. For example, concerns about pollutant emissions and fuel efficiency for various types of water craft have prompted several companies to evaluate fuel cell propulsion. Solid oxide fuel cells that are as small as a store-bought power generator could replace traditional generators in recreational vehicles and for emergency or off-grid power in homes and businesses. Scaled up, they can power hospitals, residential complexes, and other mid-size structures and co-produce electricity through central generation power plants that integrate fuel cells into their systems. Solid oxide fuel cells are easier to camouflage for military operations, too, because they are nearly silent and give off low infrared radiation, or heat.

However, for fuel cells to gain broad market acceptance, they must demonstrate their viability. With anti-idling legislation being enacted across the country, the auxiliary power unit market for diesel truck transport may become that proving ground.

Engine idling provides energy in a truck's cab overnight, but also consumes over 1 billion gallons of diesel fuel annually and releases a high amount of CO_2, nitrous oxide, and particulate matter. One option to combat greenhouse gas emissions while providing light and HVAC for resting truckers is the installation of separate auxiliary power units in the cabs of these rigs. Pyrochlore catalysts used in conjunction with hydrogen-based fuel cell auxiliary power systems not only reduce emissions, they also decrease fuel consumption and extend engine life, thereby reducing the economic and environmental costs of diesel engine idling.

NETL is now working together with industry to get the catalysts to market and has licensed the technology to Pyrochem Catalyst Company, a start-up business headquartered in southwestern Pennsylvania. The agreement marks the first time that an NETL-licensed technology has been used as a basis for the creation of a start-up company. The effort exemplifies NETL's commitment to transferring advanced energy technologies from the laboratory into the marketplace and has earned NETL recognition from the Federal Laboratory Consortium (FLC). In 2011, the FLC presented the Lab with its Award for Excellence in Technology Transfer, which honors outstanding work in transferring federally developed technology.

NETL is hopeful that the successful commercialization of the catalyst will lead to the creation of high-technology jobs in the region. Work at Pyrochem Catalyst is initially focused on further developing NETL's pyrochlore catalyst for use in fuel cell auxiliary power systems to provide non-propulsion power for vehicles, including long-haul truck transport, and to supply power in several military power applications. And, the catalyst could benefit many other industries as well.

With a technology transfer award under its belt and many possible applications, the pyrochlore catalyst holds much promise for the future of the energy industry.

Pioneering NETL Effort Applies Computational Methods to Biomaterials

With physical properties (density, elasticity, strength, toughness, etc.) that more closely represent natural bone than do other implant metallic materials (stainless steel, titanium, and cobalt-chromium based alloys), magnesium has potential for use in biodegradable stents and orthopedic fasteners that would make surgical removal unnecessary. Biodegradation of magnesium results in formation of a nontoxic hydroxide; however, the associated release of hydrogen gas would be harmful unless controlled. As a first step in developing a corrosion control strategy, researchers at NETL and the University of Pittsburgh have applied their skills and experience with calculations based on phase diagrams (CALPHAD) to investigate the thermodynamic

A nanometer is one billionth of a meter, one hundred thousand times smaller than the width of a human hair. Nanotechnology is the branch of engineering that deals with manipulating the structure of products on nanometer size scales. Applications are being found for nanoproducts in engineering, chemistry, electronics, and medicine—and in the power industry. One nanoapplication NETL is investigating is the unique size-dependent properties of nanomaterials to create new catalysts that efficiently convert CO_2 emissions to chemicals such as methane.

stability of magnesium alloys in various environments, including pure water and human blood. CALPHAD results for a simple case were in qualitative agreement with first principles calculations, which become cumbersome when applied to multicomponent, multiphase, and multi-ion systems. The details of the study are found in the December 2011 issue of the Elsevier publications, *Materials Science and Engineering B* (Vol. 176, No. 2).

NETL Research Generates Interest Within Titanium Community

NETL researchers have formulated a numerical model that predicts the compacted density of titanium powders (and perhaps other powders) as a function of pressure much better than previous efforts. An improved understanding of how compaction pressure affects density will permit better control of residual porosity and avoid excessive pressures that can accelerate die wear in powder-metallurgy processes. Described in the May 2011 issue of *Metallurgical and Materials Transactions A* (Vol. 42, No. 5, pp. 1325–1333), the model was applied to a new International Titanium Powder process that has the potential for greatly reducing the cost of manufacturing titanium components and has already received considerable interest from General Electric Co. and Boeing Co.

NETL Helps India Avoid Emitting Millions of Tons of Greenhouse Gas

A recently completed independent evaluation of the now-concluded Greenhouse Gas Pollution Prevention (GEP) project revealed that efficiency improvement activities supported by NETL at coal-fired power plants in India had avoided more than 90 million metric tons of CO_2 emissions by the end of fiscal year 2010. Further, by the end of fiscal year 2011, more than 100 million metric tons of CO_2 had been avoided. Funded by the India Mission of the U.S. Agency for International Development (USAID/India), the project's estimated cost-effectiveness, considering funding from both USAID's and India's investment, is less than $0.50 per metric ton of CO_2—far lower than alternative technology approaches. NETL provided workshops, exchange visits, and assistance in establishing the internationally recognized and award-winning Centre for Power Efficiency and Environmental Protection. Over the past 16 years, NETL scientists and engineers have given technical and management support to identify and demonstrate state-of-the-art technologies, software, diagnostic tools, testing techniques, and best operation and maintenance practices for existing coal-fired power plants in India. GEP project results will underpin the coal component of USAID's new Partnership to Advance Clean Energy-Deployment (PACE-D) as a cost-efficient way to address climate change. Under PACE-D, a private sector consultant will develop a model coal-fired power plant concept for India, provide additional training, and continue to promote U.S. technology.

Gonzaga University Offers Engineering Program in Electric Utility Transmission and Distribution

The School of Engineering and Applied Science at Gonzaga University in Spokane, WA, has developed a series of courses designed to provide working utility engineering professionals with advanced engineering instruction in modern design practice with an industry-focused theoretical foundation. Twelve online courses have been developed to date, and successful completion of any five will earn a Transmission and Distribution (T&D) certificate. Each course is developed and taught by a team of experienced engineers—experts from public and private utilities, consultants, and industry suppliers—providing students a unique opportunity to interact directly with

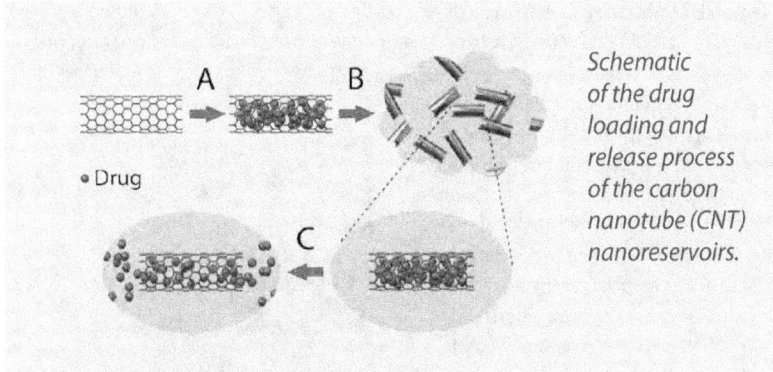

Schematic of the drug loading and release process of the carbon nanotube (CNT) nanoreservoirs.

- Drug

A. *Drug solution is filled into the interior of acid treated CNTs through sonication.*

B. *Pyrrole is added to the suspension containing CNTs and Dex and electropolymerization is carried out.*

C. *Drug is released from CNT nanoreservoirs to surroundings through diffusion or electric stimulation.*

Novel NETL Nanotechnology Supports Neural-Controlled Prosthetics—Development of neural-controlled prosthetics, which allow recipients to manipulate their artificial limbs by means of microelectrodes implanted in the brain or other neural tissue, has been hampered by performance degradation over time caused by rejection of the microelectrode implant, formation of scar tissue at the neural implant, or loss of neural tissue in areas surrounding the implant. Now, a team of scientists from NETL and the University of Pittsburgh's Bioengineering Department has demonstrated the use of multi-wall carbon nanotubes as nanoreservoirs for the loading and controlled release of anti-inflammatory and anti-rejection medicines at the implantation site. The carbon nanotube-based system outperformed standard drug delivery systems by improving the amount of drug released during electrical stimulation and increasing the lifetime of the drug delivery device. The work is described in the very high-impact Elsevier journal, *Biomaterials*, in the September 2011 issue (Vol. 32, No. 26).

different industry experts over the eight weeks of each course. Over the past three years, the T&D program has instructed more than 170 utility engineers in advanced engineering and has contributed to the training of more than 300 power system operators. This Congressionally Directed project is funded by the Office of Electricity Delivery and Energy Reliability and managed by NETL to help train the next generation of engineers in the cost-effective design, construction, operation, and maintenance of modern electrical transmission and distribution systems.

NETL Researchers Publish Book on Fuel Processing for Fuel Cells—A new Elsevier title, *Fuel Cells: Technologies for Fuel Processing*, edited by NETL researchers Dushyant Shekhawat and David Berry along with Louisiana State University Professor James Spivey, is available in both hardcover and Kindle versions. The chapter authors are globally recognized authorities in industry, academia, government research laboratories, and foreign institutes, and several are affiliated with NETL. Despite the increasing technical and commercial importance of fuel cells, few books have comprehensively addressed the critical subject of fuel reforming technology. This publication provides an overview of the most important aspects of fuel reforming to researchers, technologists, teachers, students, engineers, and interested readers. The coverage includes all aspects of fuel reforming: fundamental chemistry, modes of reforming, catalysts types, catalyst deactivation, fuel desulfurization, reaction engineering, thermodynamics, heat and mass transfer, system design, and recent research and development. The new volume will serve as an excellent self-instruction book for those new to fuel cells or as a comprehensive resource for experts in the area of fuel processing. The material is presented in a manner that also makes it suitable for referencing by graduate-level courses, fuel cell developers, and fuel cell researchers.

NO_x Management Technology Goes Commercial—Fossil Energy Research Corporation of Laguna Hills, CA, received the first commercial purchase order for a catalyst activity-monitoring system developed under DOE's Innovations for Existing Power Plants program. The non-intrusive technology provides a real-time view of catalyst deactivation within selective catalytic reduction (SCR) units used to control nitrogen oxides from the combustion of coal. Developed in partnership with NETL, Southern Company, and the Electric Power Research Institute, the Knoxcheck® Online Catalyst Activity Test system enables optimized catalyst management for year-round SCR operation by predicting remaining catalyst life and evaluating catalyst replacement options without requiring a unit outage to obtain and analyze catalyst samples.

The landmark Mt. Elbert gas hydrate field evaluation program is headquartered at the Mt. Elbert well on the Alaskan North Slope.

NETL Scientists Co-edit Special Volume on Mount Elbert Gas Hydrate Test-Well Results

—The February 2011 Elsevier Journal *Marine and Petroleum Geology* (Vol. 28, No. 2) includes 23 peer-reviewed papers on scientific findings derived from the well log, core, and pressure test data acquired at the Mount Elbert well on the Alaska North Slope in early 2007. Compiled and edited by NETL researchers Ray Boswell and Brian Anderson with Timothy Collett of the U.S. Geological Survey and Robert Hunter of the Arctic Slope Regional Corporation—a leading oil and gas services company based in Anchorage, AK—the volume includes contributions from 60 scientists representing 20 organizations in 3 countries. NETL scientists are contributors to 11 of the papers and lead authors for 4. The Mount Elbert program, supported by the Office of Fossil Energy, represents a landmark in gas hydrate field evaluation. It provides for the validation of seismic and log analysis techniques, advances gas hydrate numerical simulation, and guides selection of the most appropriate test sites for planned field-testing.

New NETL Report Compares Barnett and Marcellus Shales

—A review of the published literature indicates that the Barnett and Marcellus shales share a number of critical geologic and engineering features, but they also exhibit several important differences. Thorough knowledge of key characteristics of the two formations—such as similarities and dissimilarities in rock type, hydrocarbon potential, response to stimulation, and sweet spots—will aid operators in both the Fort Worth and Appalachian Basins. Available as a report from the publications link on the Oil & Natural Gas Supply Technologies page of the NETL website, the report will provide producers in both regions a better understanding of the controls on gas volumes and distribution, the reservoir properties that influence gas production, and drilling and stimulation techniques that can be made more effective in the future.

Patented Process Detects Mercury in Flue Gases

—NETL researchers were awarded U.S. patent 8,069,703 for a spin-off technology from their previously patented GP-254 Process (U.S. patent 6,576,092). The GP-254 Process uses ultraviolet light to photo-oxidize mercury into a more readily captured form and remove it from flue gas. The new technology provides a means of semi-continuous mercury detection by using ultraviolet light to photo-oxidize mercury and deposit it on the uncoated substrate surface of a surface acoustic wave mass sensor. When an array of surface acoustic wave sensors is employed, each sensor monitors the emission of a particular heavy metal or metalloid, such as mercury. This technology answers the need for an inexpensive and reliable method of mercury detection that can verify high levels of removal from many power plant and other gas streams, including low-rank coal-derived flue gases, incinerator flue gases, high-sulfur trioxide flue gases, and coal-derived fuel gas.

Elsevier Publishes NETL Review on Oil Shale Processing Techniques

—NETL and University of Pittsburgh collaborators completed a critical review on the dielectric properties and electromagnetic heating of Green River Formation oil shales. Electromagnetic techniques are of interest because the applied energy can selectively detect and heat organic layers present within oil shale. Published in the January 2011 issue of *Fuel Processing Technology* (Vol. 92, No. 1), the review presents and synthesizes prior work concerning the influence of applied frequency, oil shale grade, water, and temperature on the dielectric properties of oil shales and can aid future development of frequency- and temperature-specific in situ retorting technologies and oil shale grade assay tools.

National Laboratory Collaborators Co-edit Engineering Science Reference

—Researchers at NETL and Oak Ridge National Laboratory have co-edited a 13-chapter volume on the new theories, numerical methods, and applications that have emerged during the last two decades to describe gas-solids flow. The

book, *Computational Gas-Solids Flows and Reacting Systems: Theory, Methods, and Practice*, is based on a collaboration of scientists Sreekanth Pannala of Oak Ridge National Lab and NETL's Madhava Syamlal and Thomas J. O'Brien. Co-editors Syamlal and Pannala authored a chapter on multiphase continuum formulation for gas-solids reacting flows, and NETL scientist Ronald W. Breault authored a chapter on mass and heat transfer modeling.

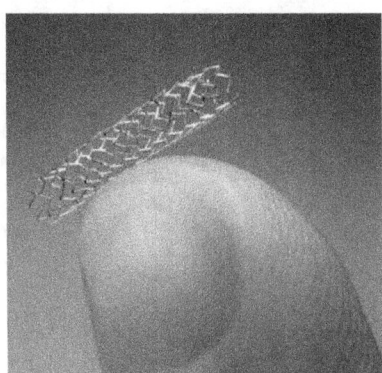

Coronary stents made from a novel platinum-chromium alloy are more flexible and conformable than traditional stainless steel stents. The novel alloy was developed by scientists at NETL and BSC.

State-of-the-Art Stent Sales Top $1 Billion

—A new coronary stent that incorporates an innovative metal alloy developed by scientists at NETL and Boston Scientific Corporation (BSC) has achieved more than $1 billion in sales and captured a 45 percent share of the coronary stent market. The stent's platinum-chromium alloy is the first stainless steel formulation with a significant concentration of platinum, a highly radiopaque element that makes the stent visible on x-ray and easier for coronary specialists to see during placement. After a lengthy series of clinical trials, BSC received foreign regulatory market approval in November 2009, and in early 2011 approval was received from U.S. and Canadian regulatory agencies, followed by approval from India. Recently, BSC announced that they have received regulatory approval in China, the world's largest stent market. BSC has announced that all future stents will use this innovative alloy in their manufacture.

Method for Producing Components with Internal Architectures Via Diffusion Bonding Sheets

—Microchannel technology is an emerging field of advanced chemical processing with applications in many industrial processes, including chemical synthesis and biomass and synthesis gas fuel conversion. Incorporating microchannel reactors into microreactor designs minimizes heat and mass transfer limitations allowing higher reaction rates, feedstock throughput, and conversion efficiency. Conventional vacuum hot-pressing methods for bonding laminae to form multichannel reactors are limited. NETL's patented invention (U.S. patent 7,900,811) is a novel multi-step process for the graduated diffusion bonding of laminates that forms a uniform and effective bond.

Enhancing Environmental Protection, Increasing Domestic Production

—"Stripper" wells are oil and natural gas wells that are nearly depleted but can still produce up to 10 barrels of oil or 60,000 standard cubic feet of natural gas per day. They provide about 20 percent of current U.S. oil and gas production. The Stripper Well Consortium, partially funded by NETL, has supported nearly 100 technology-driven projects since 2000. Reservoir remediation, wellbore clean-up, and surface system optimization projects have helped maximize resource recovery in stripper wells while minimizing environmental impacts and, ultimately, strengthening our nation's energy security.

Some of the technologies developed through these projects include—

- A gas-operated automatic plunger lift tool to remove fluids from stripper wells.

- A second tool for removing liquids from gas wells and gathering lines.

- A highly efficient submersible electric pump that reduces electricity costs.

- A smaller, lighter variable-capacity compressor and pump.

- A low-cost, real-time, wireless gauge for permanent or service applications.

- Pumper-well tender PDA and Smartphone software programs.

- An economical chemical delivery system that reduces corrosion and maintenance costs.

- A low-cost control box that optimizes production.

- A low-cost soil amendment technology for remediation and re-vegetation of brine-contaminated soils.

Retrofit Technology Improves Performance, Reduces Emissions—In the natural gas industry, thousands of reciprocating engines are used to produce electricity, compress and re-inject natural gas to increase oil production, or compress natural gas to feed it into gas transmission pipelines. These engines are aging, and new federal Environmental Protection Agency (EPA) emissions regulations may necessitate cost-prohibitive replacements. However, with funding from NETL, researchers at Kansas State University (KSU) and partners are designing and testing retrofit emissions control and monitoring technologies for these engines that will efficiently upgrade them while mitigating greenhouse gases—at a fraction of the cost of new engine replacements.

KSU researchers have developed a four-stroke cycle engine model and exhaust gas oxygen sensor model that will predict emissions from small engines better than current emission compliance measures, which are based on plume models for larger emission sources. The updated KSU models take into account small-engine characteristics and preferred catalytic conditions. Once validated, the models can be used in field engine control boards that can help meet new EPA emission standards by replacing outdated air fuel controllers.

Method for enhancing selectivity and recovery in the fractional flotation of flotation column particles—NETL received a patent (U.S. patent 7,992,718) for inventing a method of particle separation from a feed stream (composed of particles of varying water resistance) by injecting the feed stream directly into the froth zone of a vertical flotation column in the presence of a counter-current reflux stream. This allows the height of the feed stream injection and the reflux ratio to be varied to optimize recoveries based on existing operating conditions or other considerations. This novel method provides a high degree of particle collection with reduced carry-over of particles to the froth overflow, reduces or eliminates reliance on a clean wash-water supply, allows capture of coarse particles beyond the upper limiting size for liquid injection columns, allows capture of fine particles while mitigating the tendency of the low inertia particles to follow bubble streamlines and avoid capture, and provides other benefits over previously used methods of capture. Previously used methods of separation may only produce a high amount of low-grade product, are difficult to control under certain conditions, or require other materials to be added that must be removed at the end of the process. The current invention provides a method of particle separation that results in a higher quality product.

Thief Carbon Catalyst Oxidizes Mercury for Easier Capture—NETL researchers have received a patent (U.S. patent 8,071,500) for a catalyst that oxidizes heavy metals, such as mercury, from high-temperature gas streams generated from industrial sources. The catalyst uses partially combusted coal, termed "Thief" carbon, which can be pretreated with a halogen or left untreated in the presence of an effluent gas entrained with a halogen. Much of the mercury within flue gas is in a difficult-to-capture elemental form, but oxidized mercury is more amenable to capture. This catalyst provides a valuable cleanup technology by oxidizing elemental mercury. The oxidation process facilitates the capture of mercury and other metals by existing air pollution control devices now present at coal-burning power plants and other facilities. The Thief Carbon Catalyst has the dual advantage of high oxidation levels and high adsorption levels for halogens or halogen-containing compounds. The Thief carbon technology also yields the advantages of a longer catalyst life and concurrent lower long-term costs.

National Energy Technology Laboratory

1450 Queen Avenue SW
Albany, OR 97321-2198
541-967-5892

3610 Collins Ferry Road
P.O. Box 880
Morgantown, WV 26507-0880
304-285-4764

626 Cochrans Mill Road
P.O. Box 10940
Pittsburgh, PA 15236-0940
412-386-4687

Granite Tower, Suite 225
13131 Dairy Ashford
Sugar Land, TX 77478
281-494-2516

420 L Street
Suite 305
Anchorage, AK 99501
907-271-3618

WEBSITE
www.netl.doe.gov

CUSTOMER SERVICE
1-800-553-7681

U.S. DEPARTMENT OF
ENERGY

www.ingramcontent.com/pod-product-compliance
Lightning Source LLC
Chambersburg PA
CBHW081221170526
45165CB00009B/2906